WONDERS OF THE

OUT OF THIS WORLD

SOLAR SYSTEM

EDITORS

PROF. DR. PETER PACHNICKE AND WOLFGANG VOLZ

SCIENTIFIC ADVISOR

ULRICH KÖHLER

RUHR.2010 Kulturhauptstadt Europas

GASOMETER OBERHAUSEN

DLR Deutsches Zentrum für Luft- und Raumfahrt e.V. in der Helmholtz-Gemeinschaft

KLARTEXT

CONTENT

Scientific and artistic advisory council
Prof. Dr. Giovanni F. Bignami
Sabine Göge
Wolfgang Hess
Prof. Dr. Harald Hiesinger
Prof. Dr. Susanne Hüttemeister
Ulrich Köhler
Prof. Dr. Peter Pachnicke
Jeanette Schmitz
Prof. Dr. Tilman Spohn
Wolfgang Volz

PREFACE

Fascinating as the vast expanse of the universe may be, it also seems to us to be unfathomable. For that reason, images are more important in the fields of astronomy and planetary research than in any other discipline when it comes to helping us to understand them better.

The exhibition „Out of this World – Wonders of the Solar System" brings the heavens to Earth. It shows how distant celestial bodies come into being and disappear again. Unique images take us to valleys on Mars, to volcanoes on Jupiter's moon Io and to the rings of Saturn. Thanks to the most recent American and European space missions, we have spectacular pictures which bring our solar system closer to us. Not least, a fantastic installation on the impressive premises of the Gasometer allows us to experience the moon close-up.

I readily agreed to be patron of this exhibition, especially as it highlights an important area of research in an exemplary fashion. Research and innovation are vital to Germany's economic growth and prosperity, and thus also for its ability to meet the challenges of the future. That is why we have to continue doing all we can to ensure that our country leads the way when it comes to research and cutting-edge technology.

Space flight has a special role to play as an engine for innovation and technology. Its importance is also spotlighted in the Federal Government's High-Tech Strategy. For example, we have further increased spending on space research. Through our national space programme and our participation in the work programme of the European Space Agency, we are enhancing the technological performance and international competitiveness of Germany's space industry.

I welcome the fact that the exhibition in Oberhausen has further increased the public's interest in space research. If it were to generate enthusiasm in the natural sciences, particularly among young people, I would be especially pleased. At any rate, I am very grateful to all those who have, as it were, transformed the Gasometer into a telescope, thus making it possible for us to gain wonderful insights into fascinating worlds.

2009 is the International Year of Astronomy, the motto of which is very apt for this exhibition: „The Universe – Yours to Discover". I would like to wish all visitors both unforgettable and educational experiences on their heavenly journey of discovery.

FEDERAL CHANCELLOR DR. ANGELA MERKEL

Patron of „Out of this World – Wonders of the Solar System"

The European Mars Express space probe in 3D animation over the Tharsis volcanoes on Mars, Picture: NASA/ESA

Science and art are usually two unrelated matters. The Michelangelos and da Vincis of this world became extinct as man gave up seeking universal knowledge. If scientists and artists do enter into a dialogue today, this often takes the form of a meeting among specialists. Each side comes from its very own world, with differences in philosophy and terminology which make communication difficult.

Yet, scientists and artists have some key aspects in common. They work to overcome the boundaries of the thinkable. They also experience the mysterious, which Einstein refers to as the basic feeling that stands at the cradle of true art and science. People who have never experienced it, and who have lost their sense of wonder and discovery are, for all intents and purposes, dead, says Einstein.

Standing in awe and sharing the feeling with others is at once a temptation and a challenge. Oberhausen's Gasometer offers a perfect setting for displaying astronomical objects. A mighty industrial monument, a gigantic tin drum serving as a chamber of miracles, invites visitors to take a dive into our solar system. This idea was born in a dialogue between the Oberhausen Gasometer GMBH and the German Aerospace Center DLR while their successful 2007/2008 exhibition „The Eye of the Sky" was still ongoing – a compelling idea, given that 2009 was to be the International Year of Astronomy.

What followed was a process of mutual rapprochement. The two sides learnt that neither is art free of the restrictions and requirements of science, nor is science limited to being a barren accumulation of numbers. Think of the magic that lay in the moment when DLR scientists, whose research institute had supplied the scientific data for the moon sculpture, came and saw how the image data they had provided took the shape of a monumental copy of Earth's satellite.

Now we stand marvelling at an exhibition that combines the scientist's perspective with that of the artist. It is composed of images and objects from research with artfully displayed celestial bodies to form a single entity, taking visitors on a tour where the surfaces of remote stars lie within close reach. It is a tour of breathtaking images based on data transmitted to Earth from the depth of space. Cosmic dimensions come alive in full spatial resolution. Human thoughts and human emotions become equally inspired.

Even the most hard-headed science cannot be pursued without inspiration. How else could research scientists work on such fundamental questions as whether there is life beyond our own world, other than in a combination of empiricism, analytics, broad imagination and creativity?

This exhibition, orchestrated with courage and imagination, and based on sound knowledge gained by planetary research and astronomy, explains, in a unique manner, how the human view becomes wider and deeper at the same rate as new technologies for space research and remote exploration emerge, and how astronautics is more and more understood and implemented, not only as a research function, but also as a cultural undertaking.

Finally, what this joint masterpiece also demonstrates is that, even today, where there is a will and sufficient imagination, science and art can enter into a genuinely horizon-widening collaboration. The unusual approach chosen by the „Out of this World" exhibition will, I am certain, draw a large, enthusiastic audience. Visitors will be amazed, and feel alive in a miraculous way.

PROFESSOR DR.-ING. JOHANN-DIETRICH WÖRNER
Chairman of the Executive Board, German Aerospace Center DLR

„Philae", a comet lander built by DLR, on ESA's Rosetta mission
Photo: ESA/DLR

Saturn's moon Enceladus
Photo: NASA/JPL/ISI

ABOUT THE STARDUST WITHIN US

„Nothing is as beautiful as the moon of Wanne-Eickel ..."! Many of us will remember the song by Friedel Hensch and the Cypries from the Sixties, at least its title, after all, which has been used in regional commercials ever since. While the song's object of praise requires all one's imagination, and perhaps a favourable mindset, or at least reasonable weather conditions, there is no way of denying the fascination emanating from the moon sculpture and its ambience inside the Oberhausen Gasometer. The contemplative atmosphere of the entire space – emphasized by the fact that the Gasometer is an industrial monument – invites a comparison with pre-historic astrological sanctuaries, which were not simply sanctuaries, but rather also served as profoundly useful astronomical observatories.

One does not have to be a mystic nor, in fact, time one's haircuts with the phases of the moon, in order to appreciate the moon's very special significance for humanity. Scientists remind us that the moon has a stabilising effect on the Earth's rotational axis. For this reason, the zones of temperate, life-enhancing climates have remained in a stable position over long periods of time. This has allowed evolution to develop life continuously, up to and including humans, who have come to understand how these very circumstances affect our lives; humans who, it has to be said, also have the potential to upset these life conditions thoroughly. On our nearest neighbouring planet, Mars, temperate areas have migrated from the equator to the pole and back again within 100,000 years. This appears to have repeated itself at more or less regular intervals.

The way to the moon sculpture takes visitors through an exhibition, which on one hand deals with questions of essence and existence, growth and transformation in our world, and on the other hand displays tools that have made the cosmos accessible to humans. Our universe – realizing that there might be further universes, possibly entirely different from ours – came into being some 13 billion years ago in an incident we, perhaps a little naively, refer to as the Big Bang. What happened at this singularity of space-time is a profound mystery and the object of cosmological research. For a while yet, any theoretical answers that emerge – such as the string theory –unfortunately remain unverifiable by empirical research.

Our solar system came into being about 4.5 billion years ago from a nebula of solid matter. The process was described, correctly in principle, as early as in the 18th century by Kant and Laplace, yet a large number of mysteries concerning the finer details were left unanswered, and have remained the object of intense study ever since.

We owe the fact that we can now visualize these processes in the images shown at the exhibition to the technological developments of the past five or six decades, as well as to the apparently incessant curiosity of our species. This also applies to the entire exploration of our solar system.

The moon, as everyone knows, plays a special part in the history of astronautics. The manned lunar landing in 1969 is seen as one of the defining moments of the last century. Today, we limit our activities to sending astronauts into orbit to conduct experiments under

Keeler Crater on the Moon
Photo: NASA

Iuventae Chasma on Mars
Photo: ESA/DLR/FU Berlin (G. Neukum)

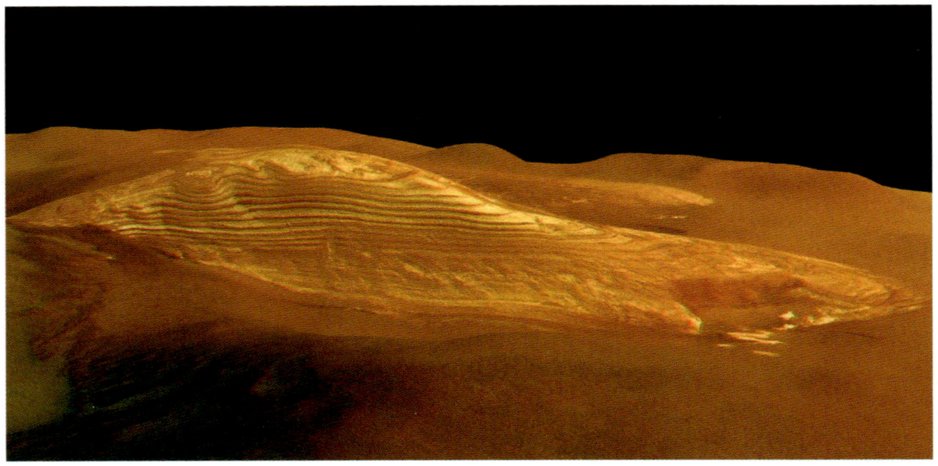

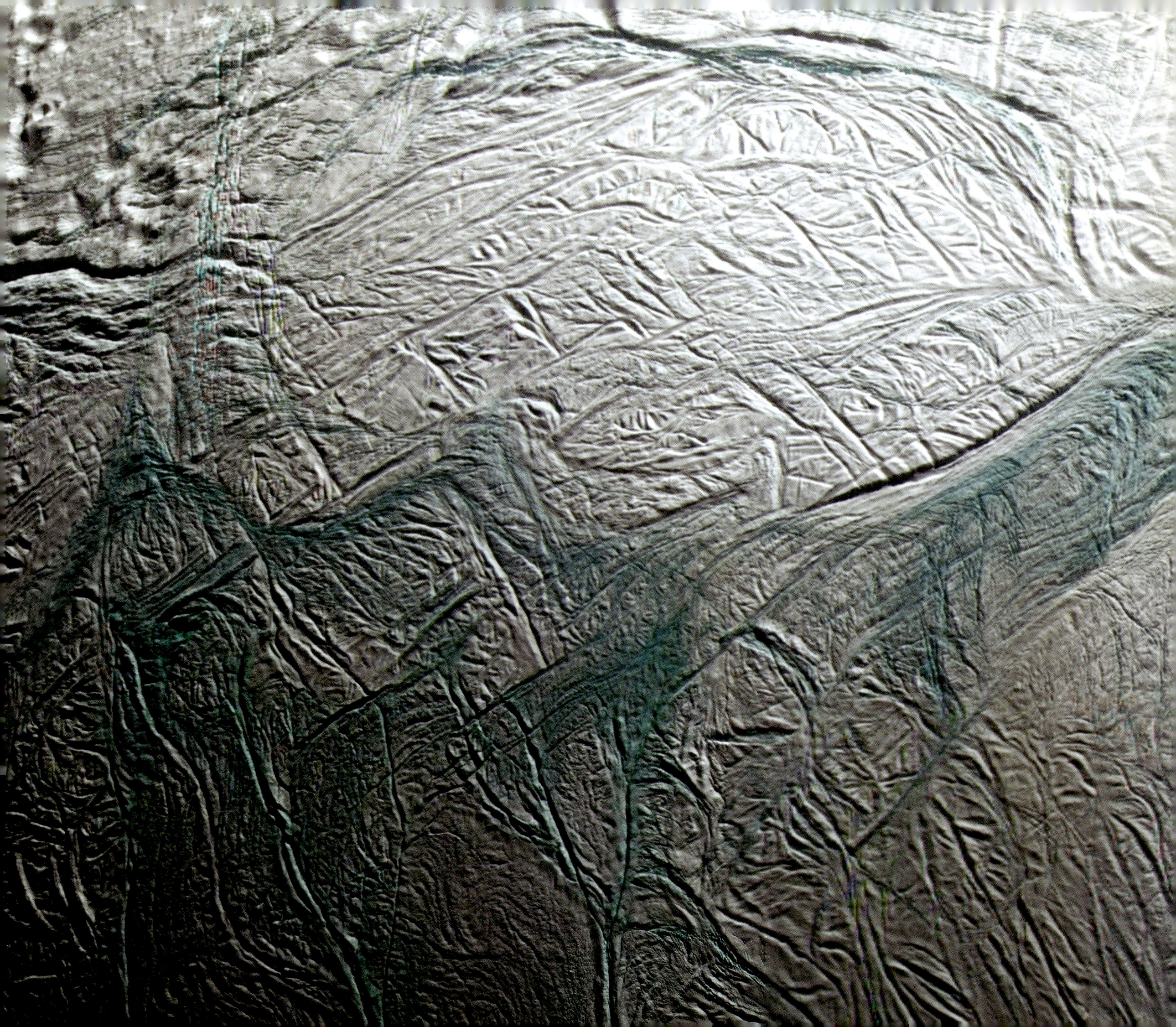

Gravitational lens in the Abell 2218 galaxy cluster
Photo: NASA/ESA/STScl/STECF

conditions approaching weightlessness. Their missions involve twofold fascination: the astronauts share their accounts of beauty of the universe, while many of us are fascinated with the astronauts themselves as modern heroes. The limitation to orbital missions does not mean that we are unable to re-visit the moon! Or, in fact, go further and visit Mars! It all comes down to a matter of will and resources.

Many people ask about the usefulness of it all. Like any other area of basic research, scientific astronautics serves to augment human knowledge, and is not required to produce an immediate benefit. And yet, space technology has become an established part of our lives. If we were to de-activate all satellites, modern traffic could no longer be managed, global communication would collapse, and the worldwide flow of commodities would be dramatically slowed. Even theories as exotic as Einstein's Theory of Relativity and quantum mechanics have found applications in technological navigation systems. Science is a driver of technology! Who knows today how we may even be using the string theory in the future?

For the moment, we are exploring our solar system with automatic probes, and we observe the universe from satellite-borne telescopes. In doing so, we regularly come up against discoveries that we would have considered impossible, and we test the limits of our technology. This is a typical phenomenon in scientific astronautics which is all about penetrating uncharted terrain. We usually have a vague idea what to expect at the other end – after all, we have to design the satellites and their instruments accordingly. And yet, things often turn out to be different. Sometimes they are quite different. Who would have expected to find dust-covered ice floes on Mars, sulphur geysers on Io, or ice geysers on Enceladus? On the other hand, how must a solar cell be designed so that it works on Mercury, or even closer to the sun, at temperatures around 300° C? How does a balloon work in Titan's atmosphere? How will instruments survive Jupiter's radiation belt? How do we determine the distance between two satellites to measure gravitational waves? How do we build the high precision clocks required? Questions that only technological progress will be able to answer.

The Herschel space telescope
Photo: ESA/STScl

The images displayed at this exhibition are overwhelmingly beautiful. The one thing that fascinates us even more is the question of whence and whither, and the question if we are really alone in the universe? The thought that humans are stardust, and that the matter we are built from is eternal, while our shape is transitory like any structure, shows us to be part of a greater entity. Assuming that this is so, there is likely to be other life beyond that on our Earth. The great physicist Enrico Ferni once called it a paradox that we have not been in touch with other intelligent civilisations, given the number of stars. Scientists have attempted to calculate the probability of the existence of another Earth, and to estimate the distance between them and us. The equation is a simple one – a product of probabilities – but we have insufficient knowledge of the size of the parameters. The distance range around the sun in which we can exist, is a narrow strip; a large moon circling a relatively small planet is another rare thing to find. We also do not really understand whether our existence is a function of plate tectonics, or even why Earth's tectonics works how it does. Did, perhaps, human evolution emerge as a result of plate tectonics?

The answers to these questions can only be found by research on Earth itself and in its surrounding space. We therefore go along with T.S. Eliot who wrote: „We shall not cease from exploration, and the end of all our exploring will be to arrive where we started and know the place for the first time."

PROF. DR. TILMAN SPOHN
Member of the Scientific Advisory Committee of the exhibition „Out of this World – Wonders of the Solar System" and head of the DLR Institute of Planetary Research

ABOUT THE EXHIBITION

For millennia, humans have gazed towards the heavenly spheres to understand how the movement of the stars might be connected to their own fate. The exhibition „Out of this World – Wonders of the Solar System" shows our solar system as a gigantic cosmic process of birth and decay. Fascinating images, spectacular exhibits, as well as historical and modern instruments of space research make the evolution of our solar system come alive. The unique architecture of the Gasometer, a cathedral-like piece of industry heritage, enables us to experience the cosmic dimension from three levels. „Out of this World" is an exhibition where the perspectives of the history of civilization, science and art merge into one, appealing equally to our intellect and to our senses.

The images of the exhibition are compellingly sensual and engaging. On the ground floor, we see monumental pictures, two by three meters in size, which convey a sense of our being an intrinsically linked to the cosmic events displayed. At first glance, it is their beauty that fascinates us. As we pause to take a deeper interest in their content, they begin to tell us their stories: about the birth of our sun and its planets, their development, their uniqueness, as well as their foreseeable deaths. We see before ourselves the entire life cycle of our solar system. Our galaxy holds billions of solar systems in highly diverse developmental stages. Astronomers have spent the last few decades recording each of these phases, using sophisticated observation and image processing techniques.

Every civilization has its own fascinating narratives on the origins of Heaven and Earth. At the middle level of the exhibition, in the compartments formed by the building's iron skeleton, we find a selection of unique historical documents and scientific instruments from past and present. These remind us of how our ideas about the origin and function of the solar system have changed over time: from the myths of the primitive peoples, to the creation myths of the world religions, finally to today's scientific view of the world. Scientific observation instruments and measuring devices demonstrate that our knowledge of the solar system took a step forward every time new technologies became available. These widened our view into both the micro and macro cosmos. 400 years ago, when Galileo looked beyond the Earth through a telescope for the first time, a process began which continues to this very day, changing our visual and perceptual capabilities. Modern telescopes and spectroscopes now permit us to record not only visible light but also the non-visible portion of the electromagnetic spectrum. Images in x-ray, ultraviolet, infrared or radio wavelengths show us not only the optical appearance of an object, but also its temperature, density, energy flows and chemical composition. Many of the images in this exhibition are a combination of shots taken at different wavelengths. These images bring those wonders of the solar system hidden to visible light before our eyes.

This technique enables us to see what might justifiably be called the greatest of all miracles: the spectacle of the birth, development, death and reincarnation of stars. What this process produces is not recurring sameness; rather, it is a continuous metamorphosis. Over billions of years, the creations of new and ever more complex molecular compounds, up to and including organic life forms, have been born from the dust of dying stars. On our Earth, this fascinating evolution has produced us humans, who have the ability not only to perceive and study this colossal creative process of the universe, but also to enjoy its overwhelming beauty with all their senses.

„It is no exaggeration to say that humankind is stardust, since the iron in our blood, the calcium in our bones and all the other heavy metals in our bodies have originated in the explosion of a supernova from which our solar system and, finally, ourselves have evolved." (Michael Shara)

PROFESSOR DR PETER PACHNICKE, WOLFGANG VOLZ

STARBIRTHS

In the life of the cosmos, the birth, development, death and rebirth of stars form an endless cycle in which the dust and gas left behind by dying stars provide the raw material from which new stars are made. Dust and gas contract in enormous clouds within which, falling inward under the gigantic pressure of gravity, they spontaneously ignite the nuclear fire that makes a star shine.

In recent years, the Hubble space telescope has provided us with impressive and spectacular images of sites where stars are being born in our galaxy. Because of their dramatic character, the astronomers call these sites star-birth fields, stellar cradles or breeding grounds. They hatch not only individual stars but entire star clusters which, though formed at the same time, differ in mass, energy output and life expectancy. There, we find small brown and red dwarfs next to stars the size of our own Sun that shine for billions of years as well as giant stars that burn up all their energy in a few million years.

In visible-light images we cannot observe the process by which stars originate because it is hidden behind a shroud of dark clouds. However, Hubble's infrared camera can penetrate these protecting envelopes and show us the newborn stars within. In the Orion nebula the infrared camera detected for the first time the disks of protostellar dust which surround stars that are forming their own planetary systems. Our Sun was surrounded by such a disk when it was young. Consisting of the residual dust and gas left over from the formation of the Sun, it spawned heavenly bodies that grew larger over millions of years until they ended up as gas planets in the outer reaches and as rocky planets in the inner reaches of the solar system. Whatever material was not consumed in this process keeps circling the Sun to this day, mainly in the form of comets beyond Neptune and asteroids in a belt between Mars and Jupiter. Together with its unique system of planets and planetary moons, our Sun formed 4.5 billion years ago. Today, it has enough energy left to go on shining for another six billion years.

18_19

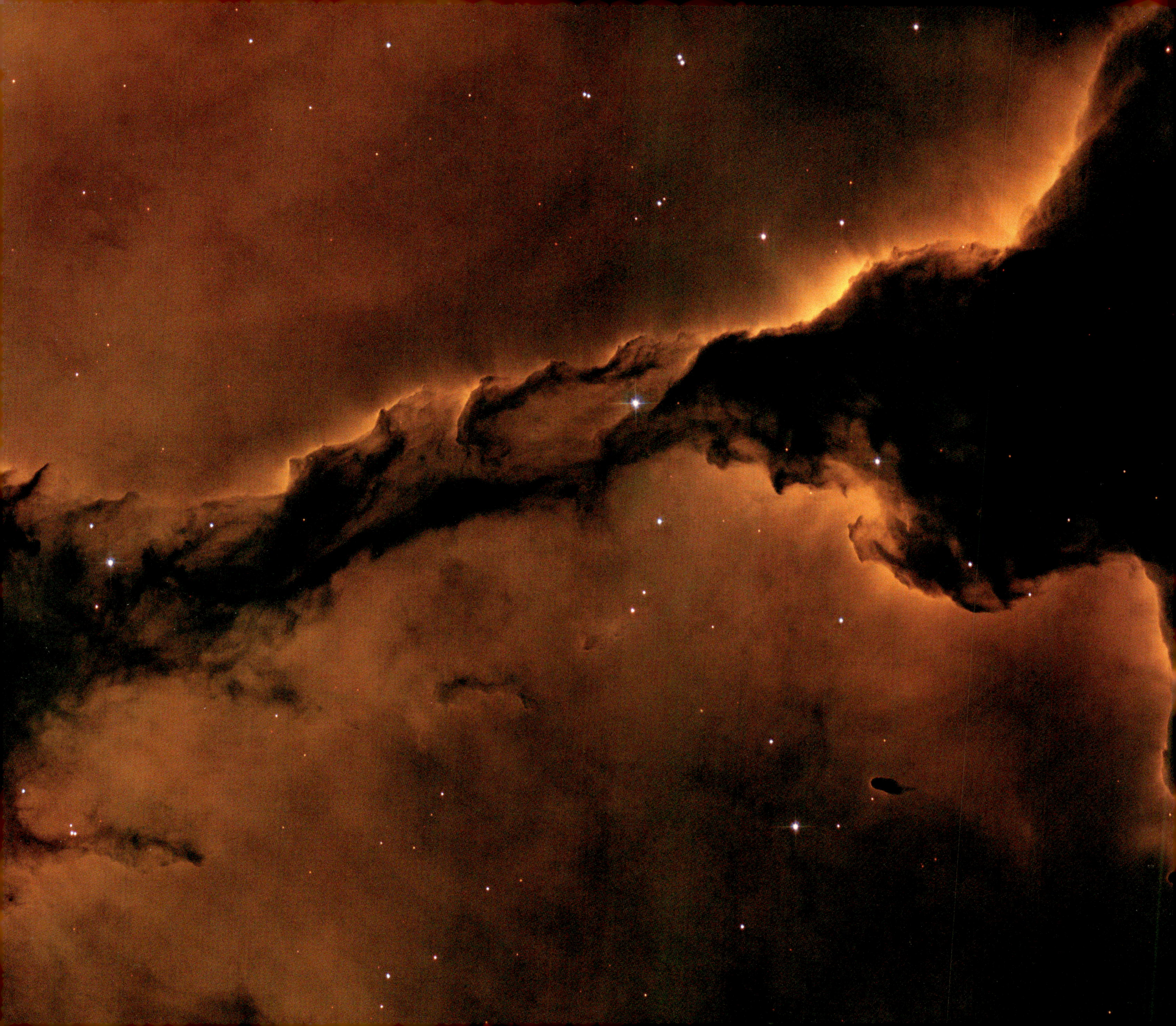

THE SUN – OUR SPRING OF LIFE

The Sun is a gigantic ball of gas, large enough to encompass more than one million planets the size of Earth. In its center, hydrogen fuses into helium at a temperature of nearly 16 million degrees in a process which releases energy in the form of radiation. Slowed down by the Sun's enormous density, this energy gradually migrates to the surface whence it is emitted into space. After another journey of 150 million kilometers which takes somewhat more than eight minutes, it reaches Earth, where life has only the Sun to thank for the light and warmth it needs.

Although the Sun shines quite steadily, explosions are continually happening on its surface. This activity fluctuates in a cycle that extends over approximately eleven years, as indicated by the dark Sunspots whose numbers increase as the Sun grows more and more explosive.

Our Sun is an average star. 4.6 billion years ago, it developed from a cloud of gas and dust together with its planets. It will go on producing energy for another 6 billion years during which its brightness will gradually increase. Thus, in no more than 1.4 billion years' time the Earth will be hot enough for the oceans to evaporate. 6 billion years hence, the Sun will have consumed the hydrogen stored in its center and expand into a red giant with a diameter 250 times greater than today. This hot ball of gas will swallow up both Mercury and Venus as it expands, while the Earth will probably travel around the outer fringes of the giant red Sun.

24_25

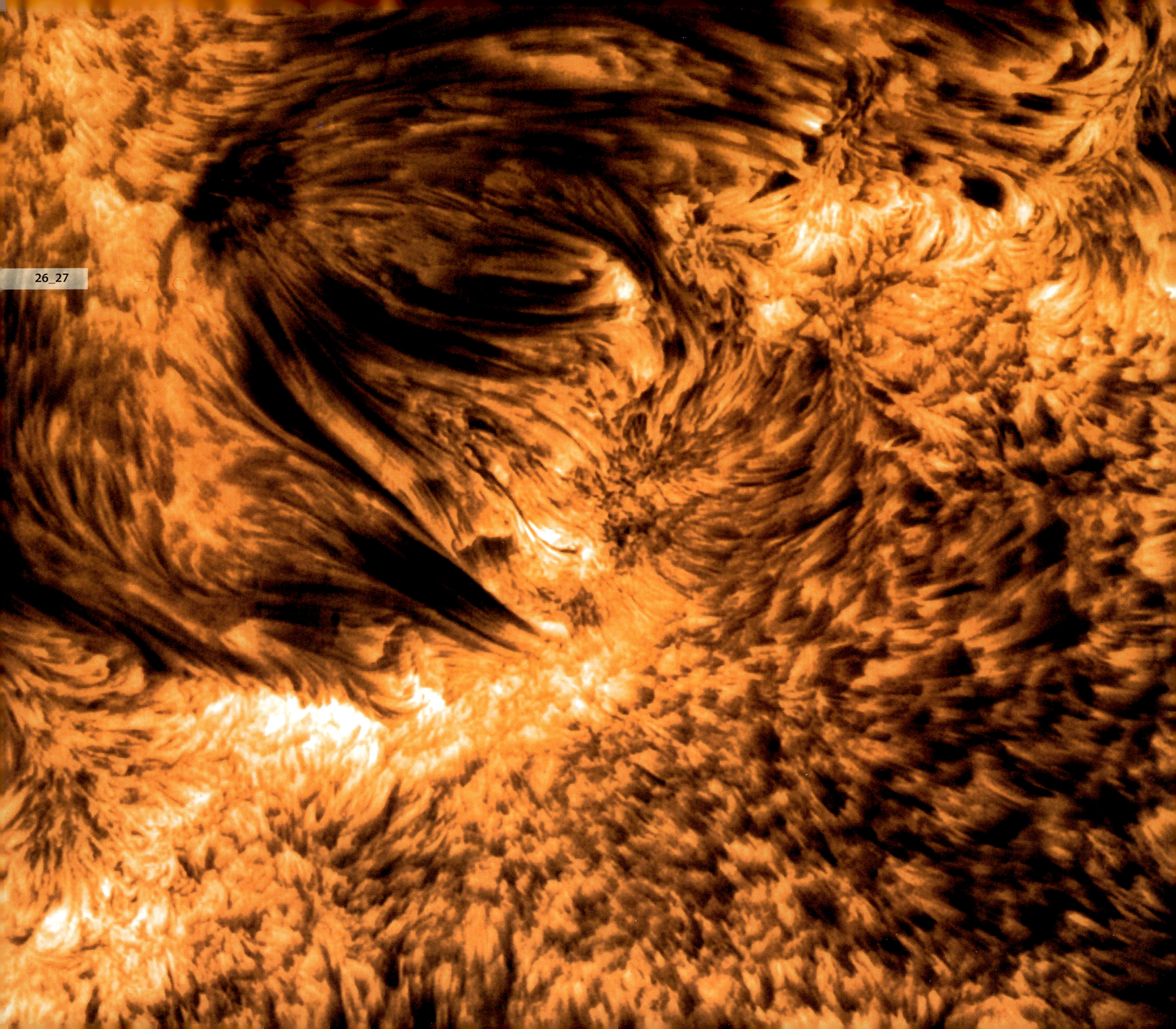
26_27

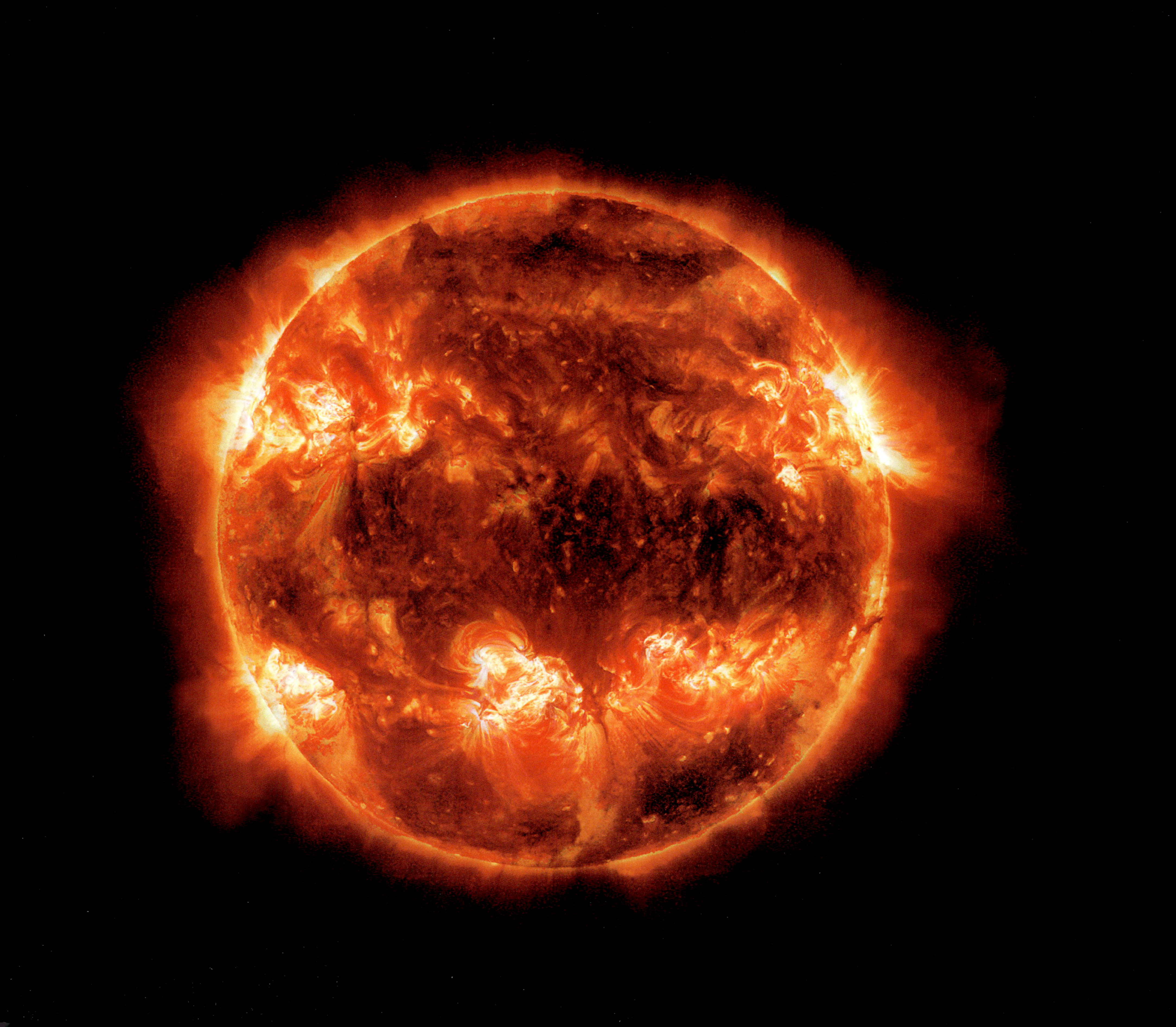

MERCURY – A DESERT OF BURNING ROCK

Mercury is the smallest planet in our solar system, and it is nearest to the Sun. Never appearing in the sky farther than 28 degrees away from our central luminary, it can only be observed close to the horizon at dawn or dusk, provided conditions are favorable. Most people have never seen it consciously. On the day side of Mercury, temperatures may rise as high as 470 degrees centigrade; lead melts at 327 degrees. Scientists suspect that the center of the planet consists of an enormous iron core that extends across three quarters of the planet's diameter, making it approximately as large as the Moon. Why Mercury of all planets should have such a large metallic core is an open question. One hypothesis suggests that it was hit by another large body during its formative phase, and that part of its rocky crust was blasted into space by the impact while the core remained intact. This could explain why Mercury's iron core today is so unusually large relative to the thickness of its crust. Being a small planet with a diameter of 4,878 kilometers, Mercury's interior should have cooled off and solidified by now, and yet it has a global magnetic field. It may be that the iron core, for unknown reasons, has remained at least partially liquid. As in the Sun and the Earth, the hot matter flowing inside Mercury, being practically equivalent to a powerful electric current, generates a magnetic field.

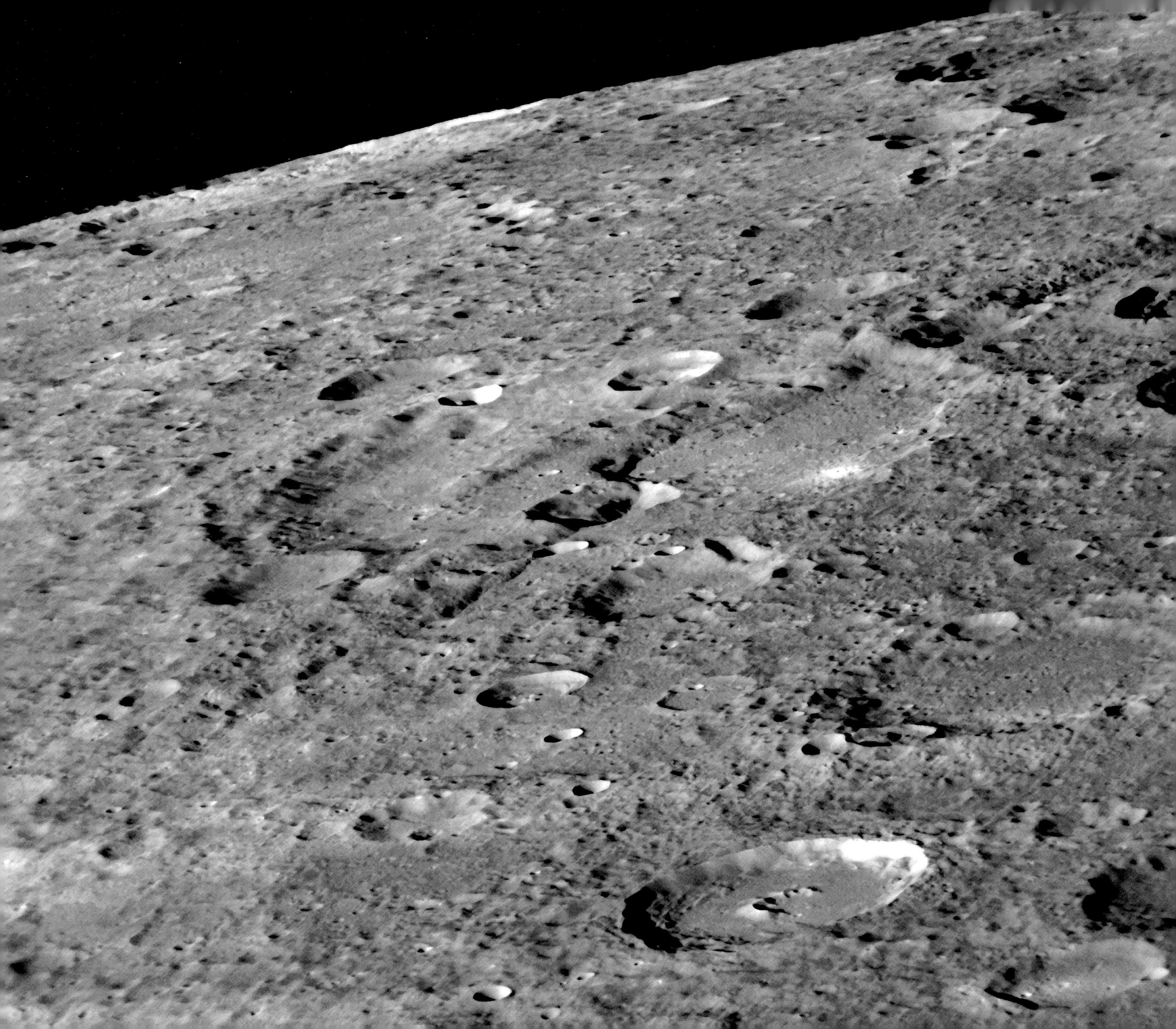

32_33

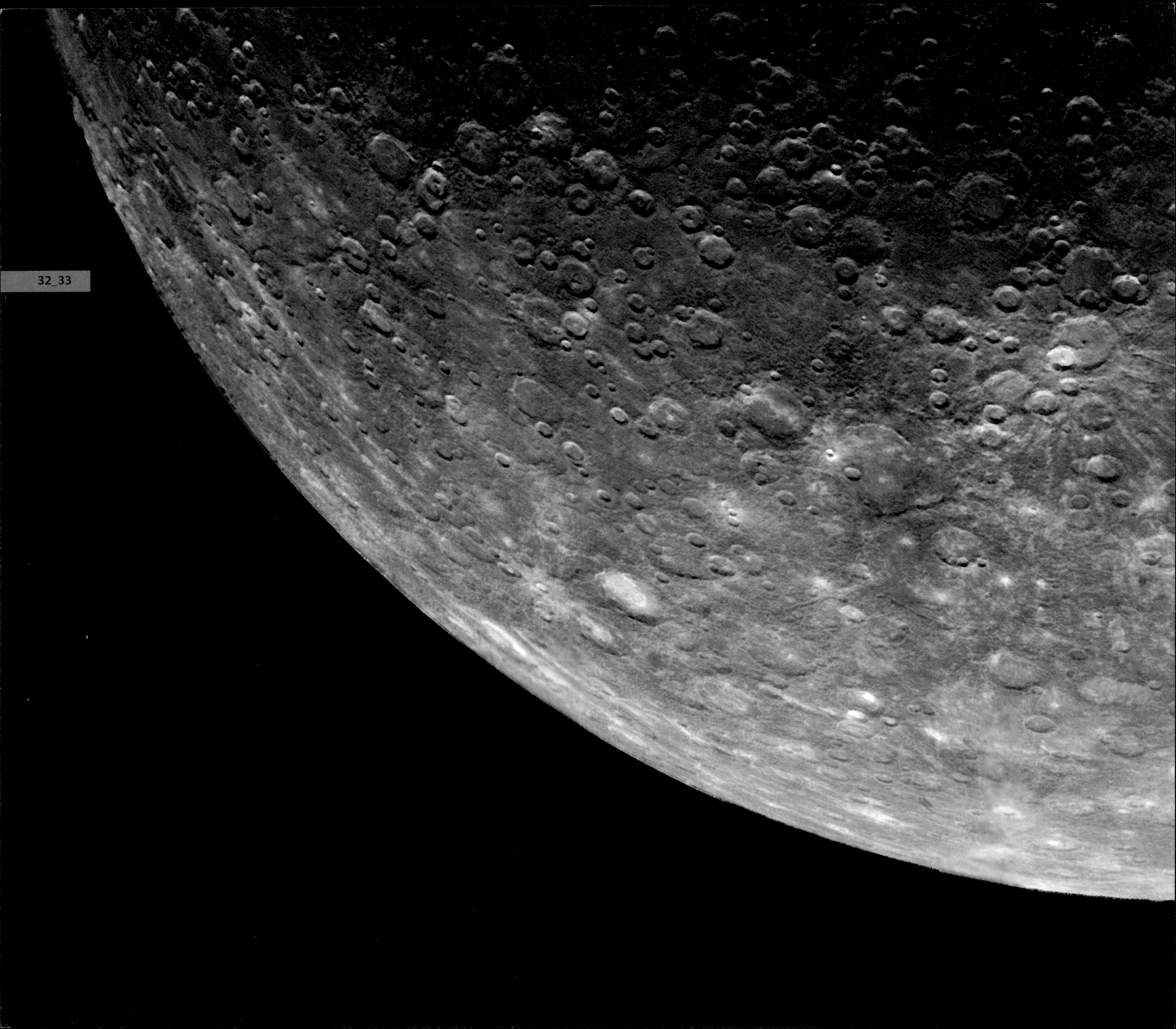

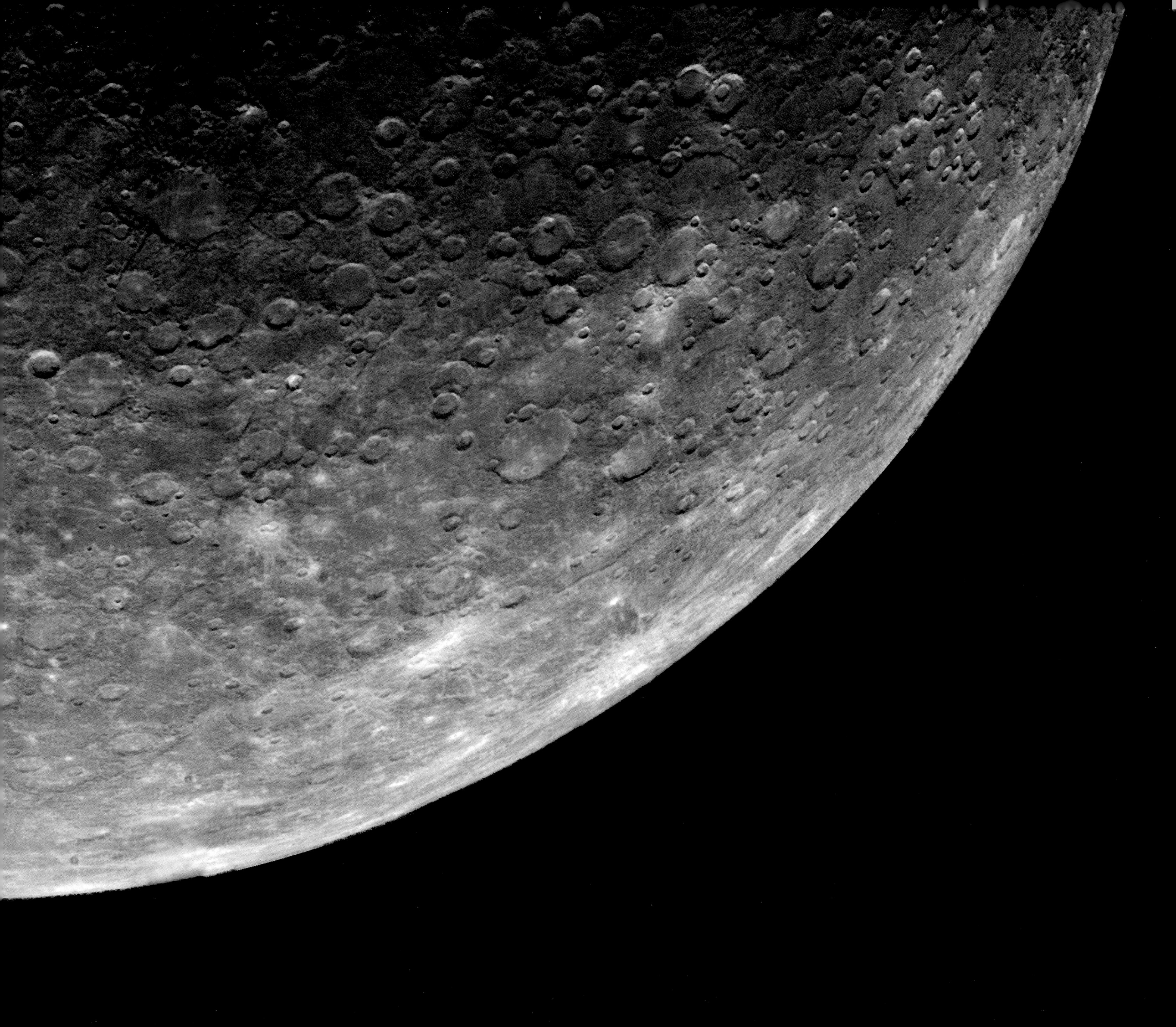

VENUS – THE SHROUDED PLANET

Although it is occasionally called Earth's twin, the similarities between Venus and our planet are soon exhausted: its inner structure resembles that of Earth, and with its diameter of 12,100 kilometers Venus is only a little smaller. Yet the temperatures that prevail on its surface range around 460 degrees centigrade and the atmospheric pressure amounts to about 92 bar, the same pressure as that in the Earth's oceans at depths of more than 900 meters.

Conditions are so extreme because the planet is close to the Sun, and because its atmosphere is composed almost exclusively of the greenhouse gas carbon dioxide and contains practically no water vapor. This is surprising because Venus and Earth were probably surrounded by similar atmospheres shortly after their formation. However, as Venus is much closer to the Sun than Earth its climate has always been hotter. Consequently, water molecules would rise into the higher strata of the atmosphere without condensing into clouds, as they do on Earth. At an altitude of about 100 kilometers, the UV light emitted by the Sun would break down the molecules of water (H_2O) into hydrogen (H) whose lighter atoms escaped into space, whereas the heavier oxygen (O) atoms returned to the surface where they were chemically bonded in the rocky material. Thus it happened that Venus dried out and turned into a searing hot desert.

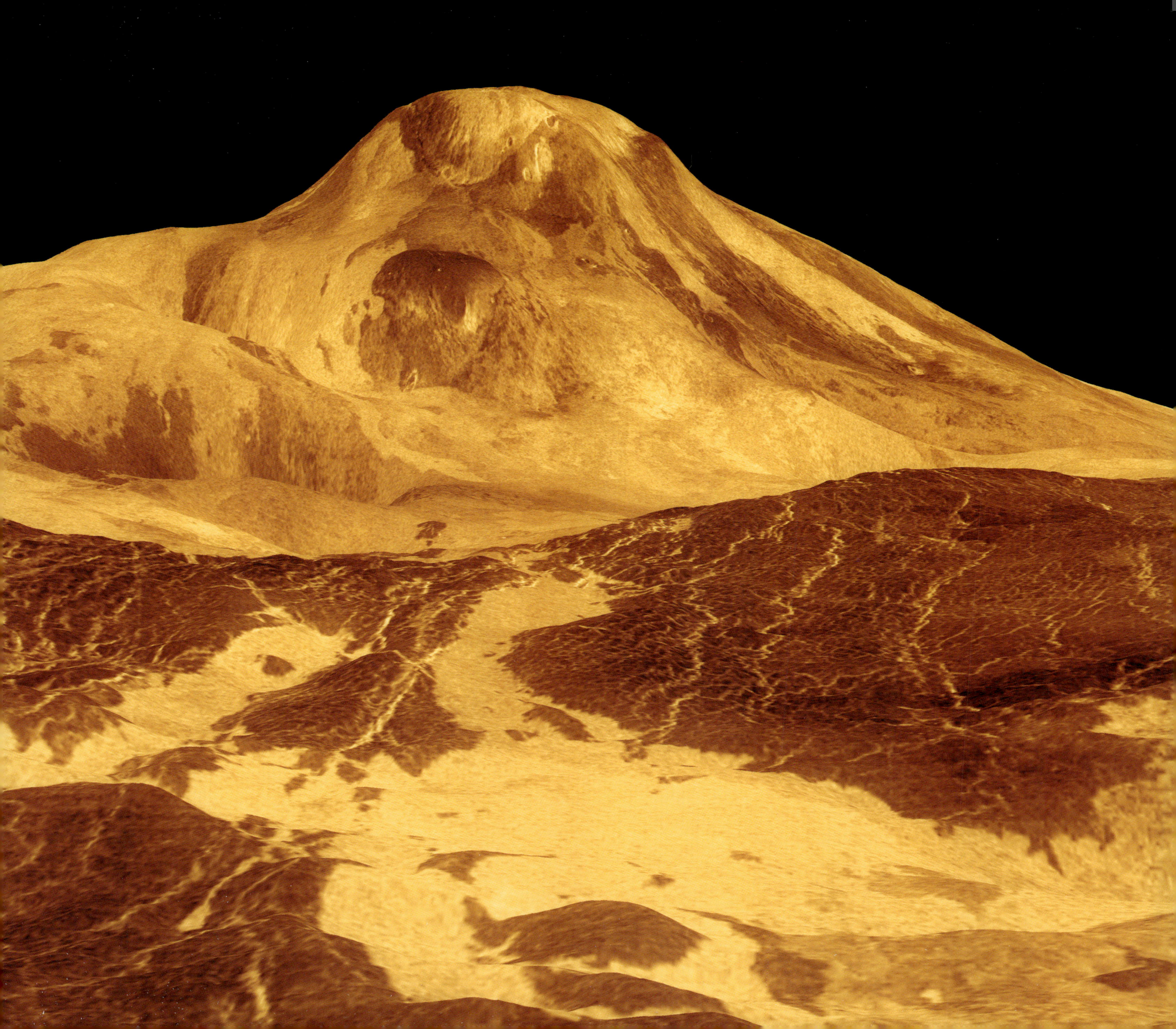

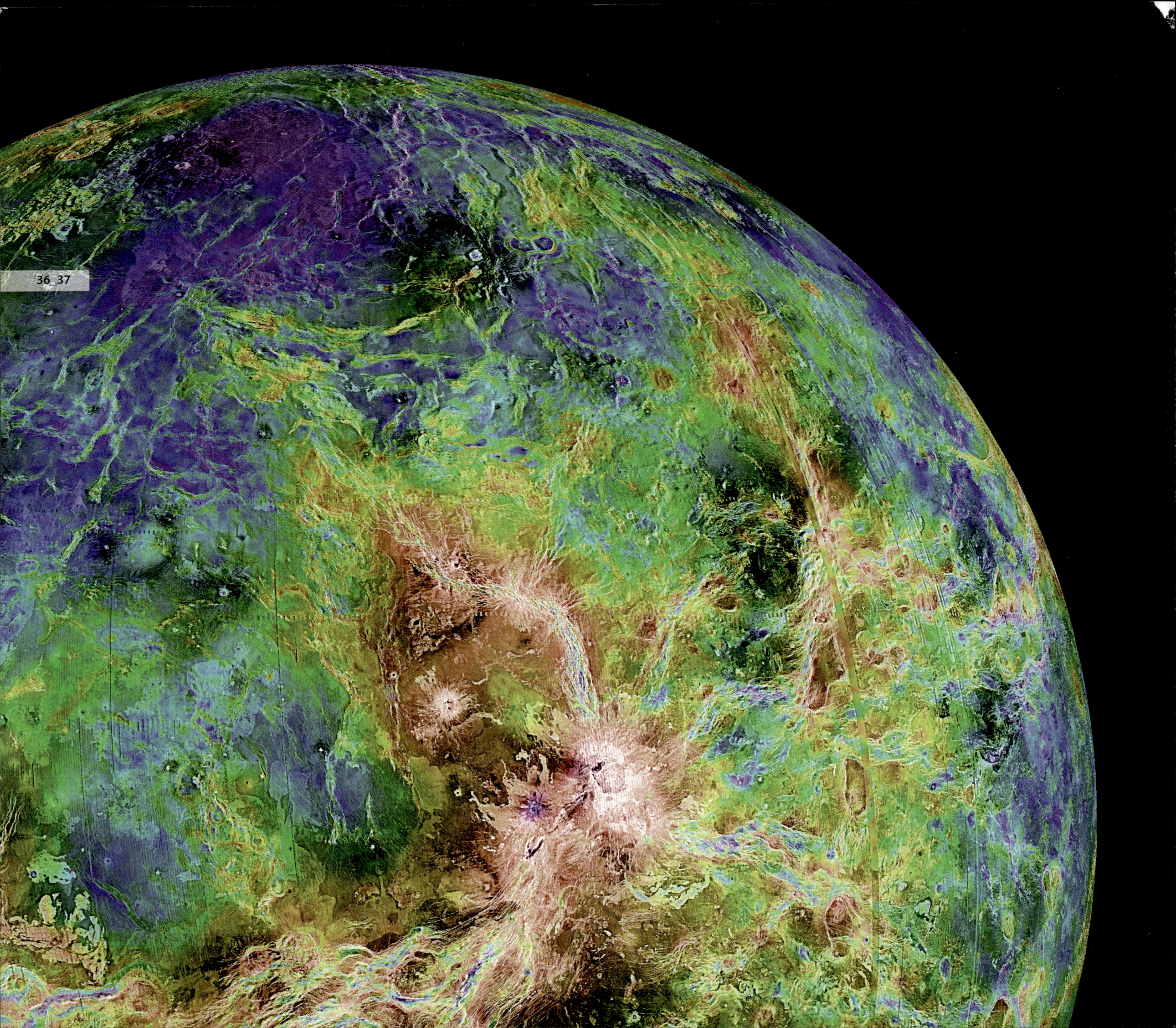

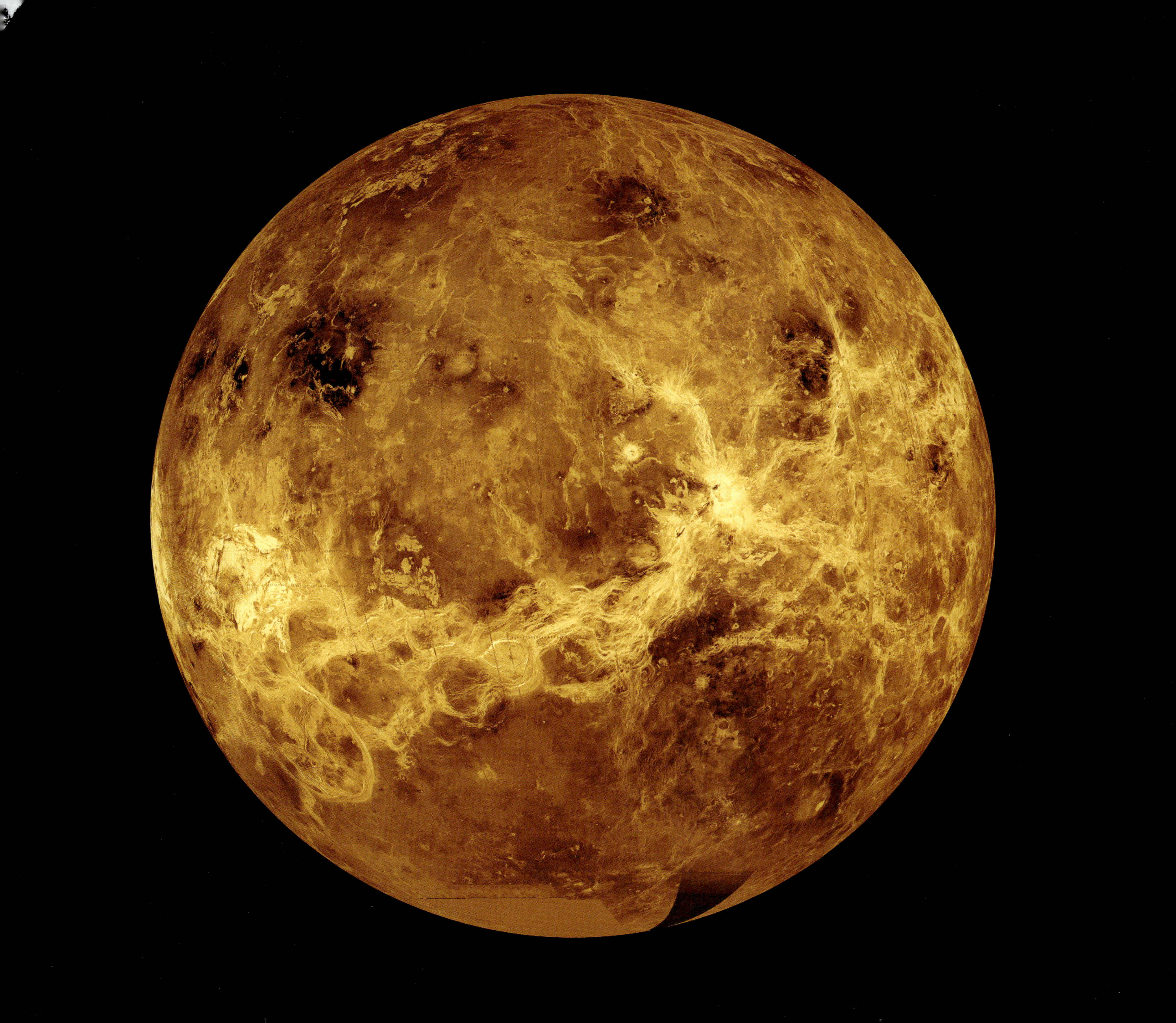

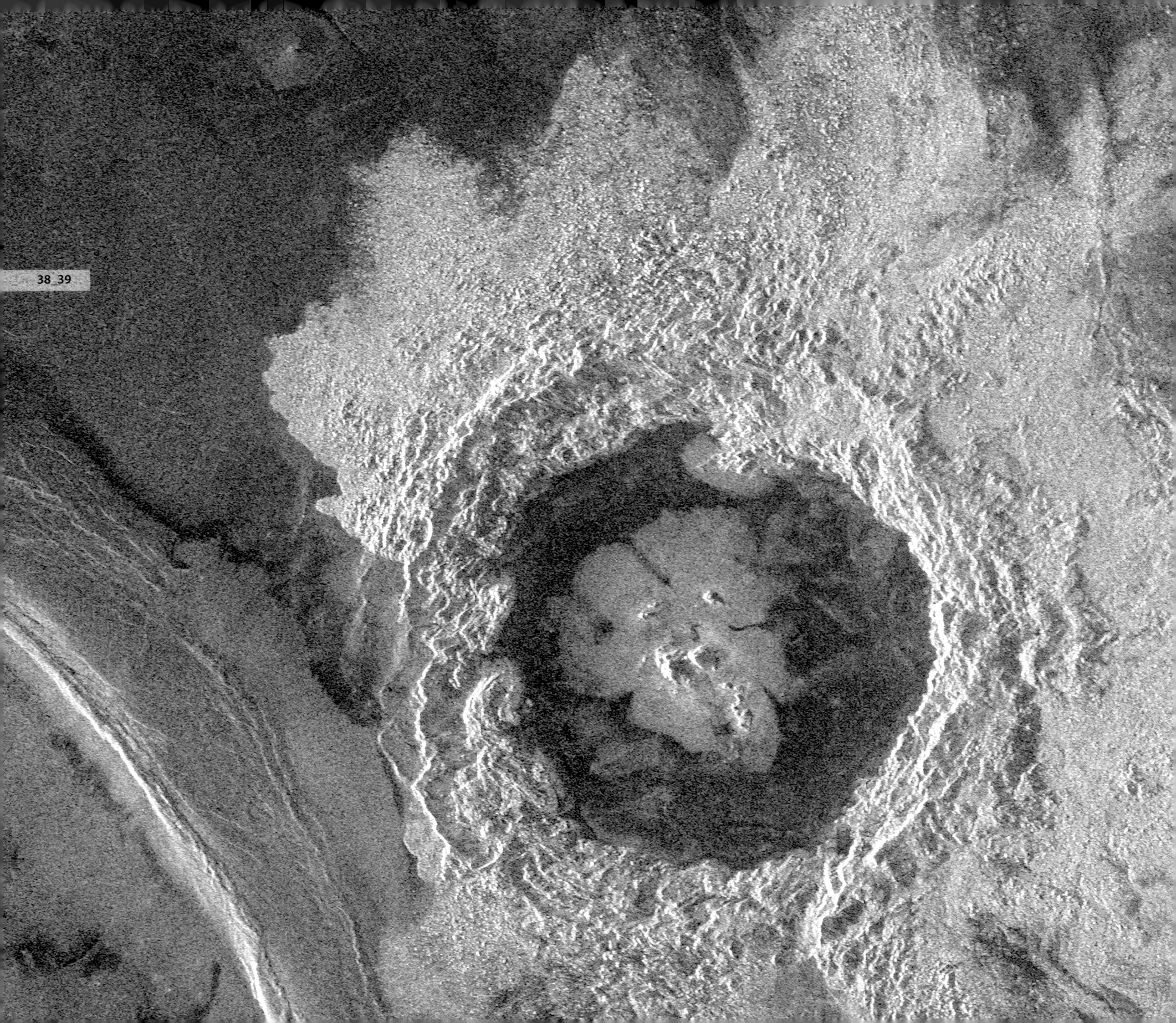
38_39

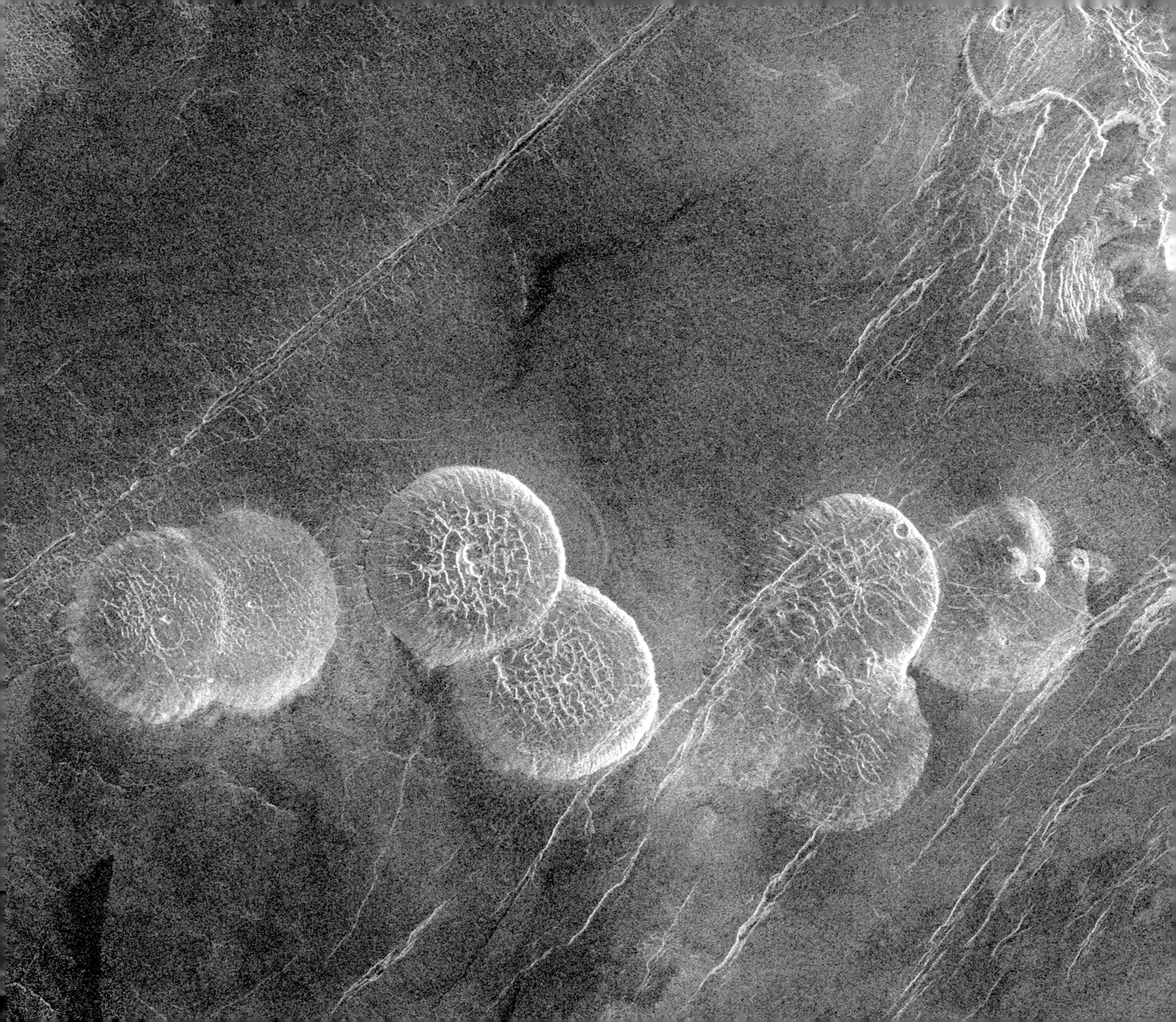

MARS – OUR RED NEIGHBOR

No other planet has exercised the imagination of mankind as much as Mars. Late in the 19th century, a great stir was caused by the discovery of what was thought to be a system of canals which, as some people supposed, had been built by a highly-developed technical civilization, an assumption that proved false in the end. Today we know that Mars is a dismal desert planet. However, this may have been different in the past. Initially, the red planet had a dense atmosphere. The climate on its surface was warm; and water was flowing through deep ravines and accumulating in lakes. It cannot even be ruled out that life may have developed at that time. Today, there still are several respects in which Mars resembles our planet. A day there lasts about 24½ hours, and seasons change, although they are twice as long as ours. In spring, the polar ice caps gradually melt, and they grow again in winter.

Mars is surrounded by a thin atmosphere consisting of 95 % carbon dioxide and 2.5 % nitrogen. Pressure at the surface amounts to somewhat less than one hundredth of that on Earth. At night, temperatures drop below minus 100 degrees centigrade, while in the daytime they may rise as high as 20 degrees centigrade in some regions during the summer. Under these conditions, it is impossible for either liquid water or – presumably – life to exist on the surface.

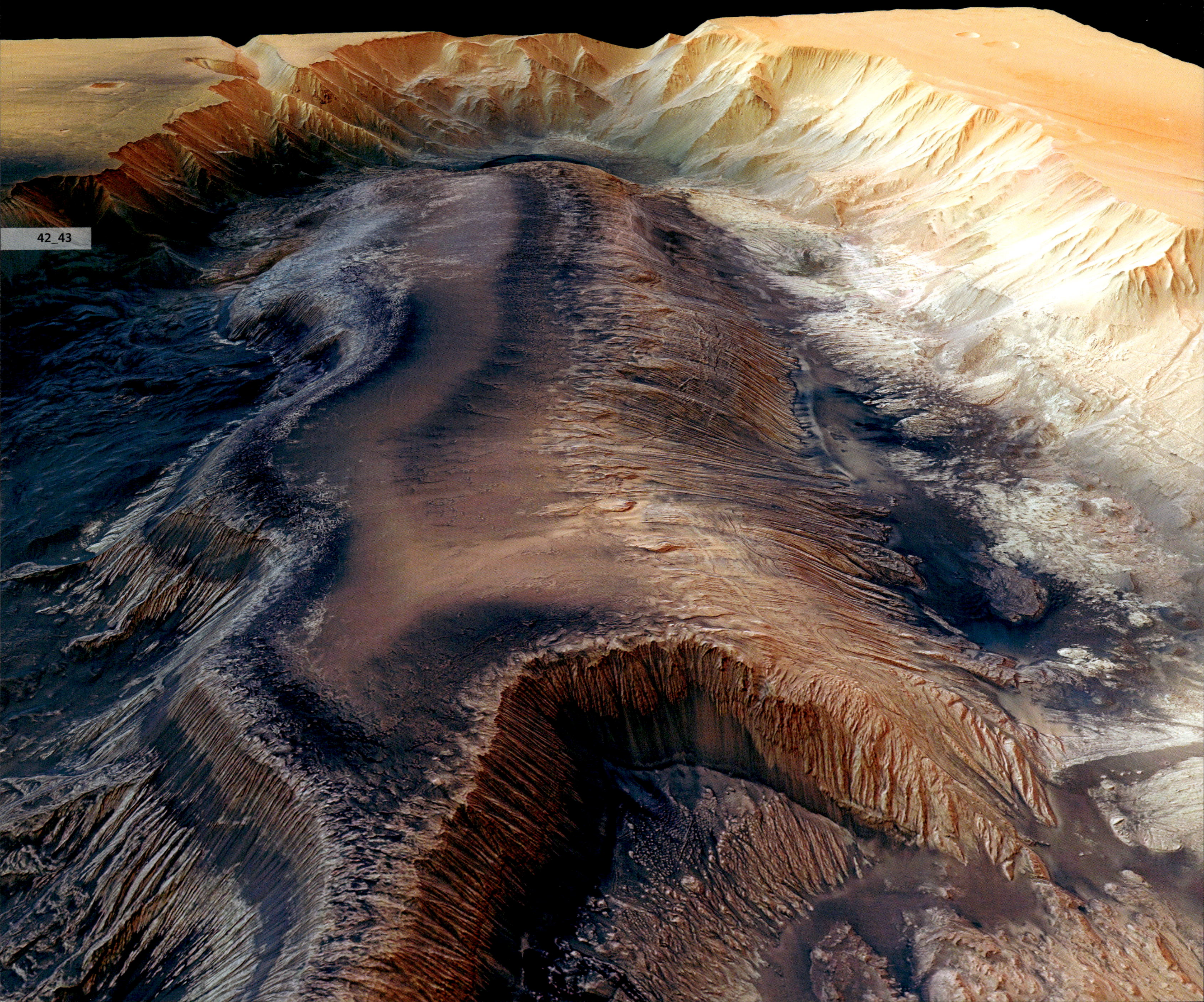

JUPITER – KING OF PLANETS

Jupiter is the biggest planet in our solar system, with space in its interior for around 1,400 Earths. More than anything else, it is its atmosphere which distinguishes it from the inner planets – Mercury, Venus, Earth and Mars – because it is more than 120,000 kilometers thick. The storms that rage in it reach peak velocities of 540 km/h. Because of its great distance from the Sun, the temperature at the upper limit of the cloud layer is only minus 150 degrees centigrade. At greater depths, temperatures and pressures increase swiftly. Around 50 kilometers down, the pressure reaches 10 bar, and room temperature prevails. Water droplets and even organic substances may probably be found at this altitude. For this reason, it has been speculated that exotic lifeforms might have evolved there. However, this appears extremely improbable in view of the severe atmospheric turbulences which researchers expect in that zone.

Scientists think that the planet's center is occupied by a rocky core about 15 times the weight of Earth. Jupiter is surrounded by rings and is known to have 63 moons.

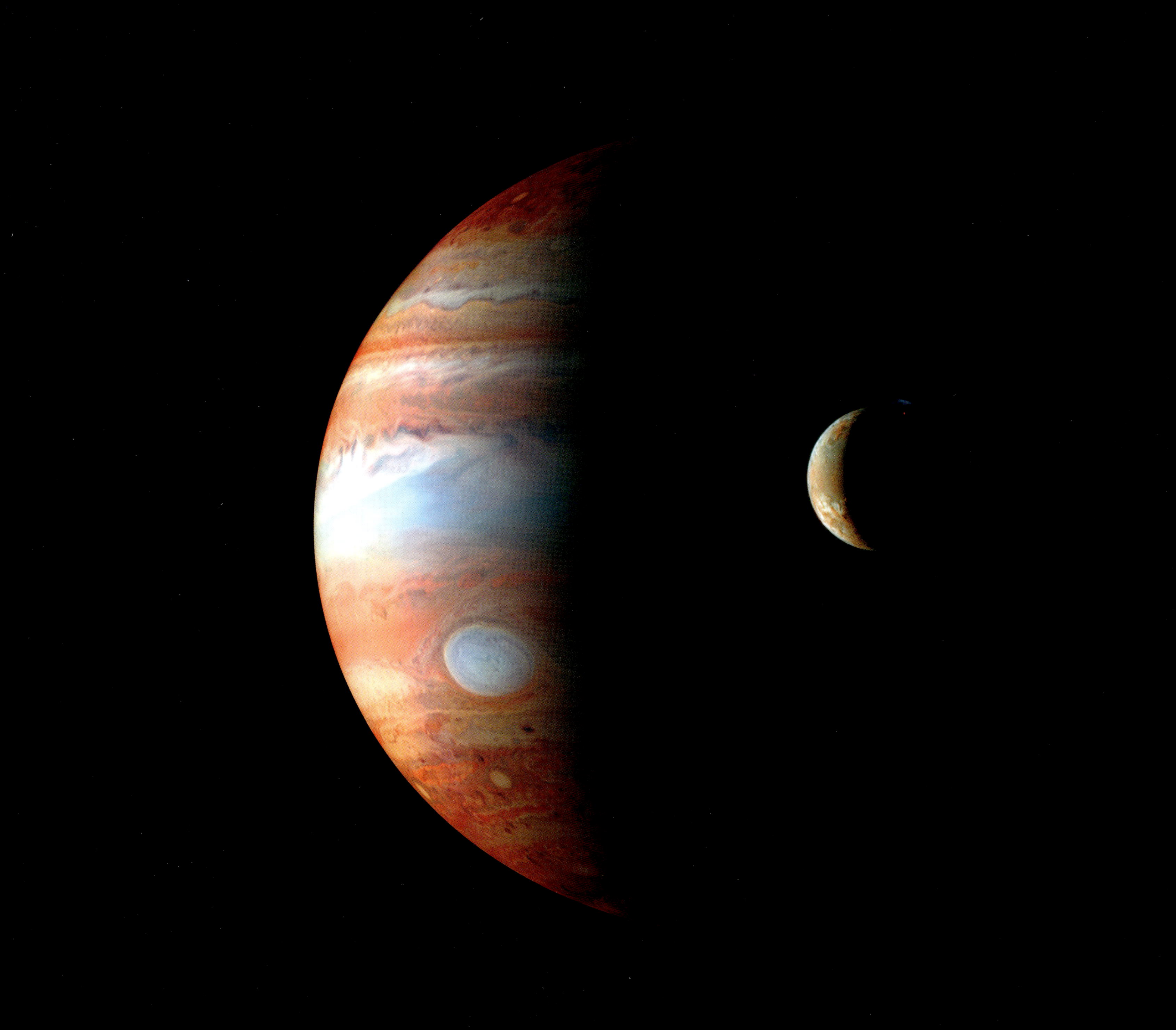

52_53

54_55

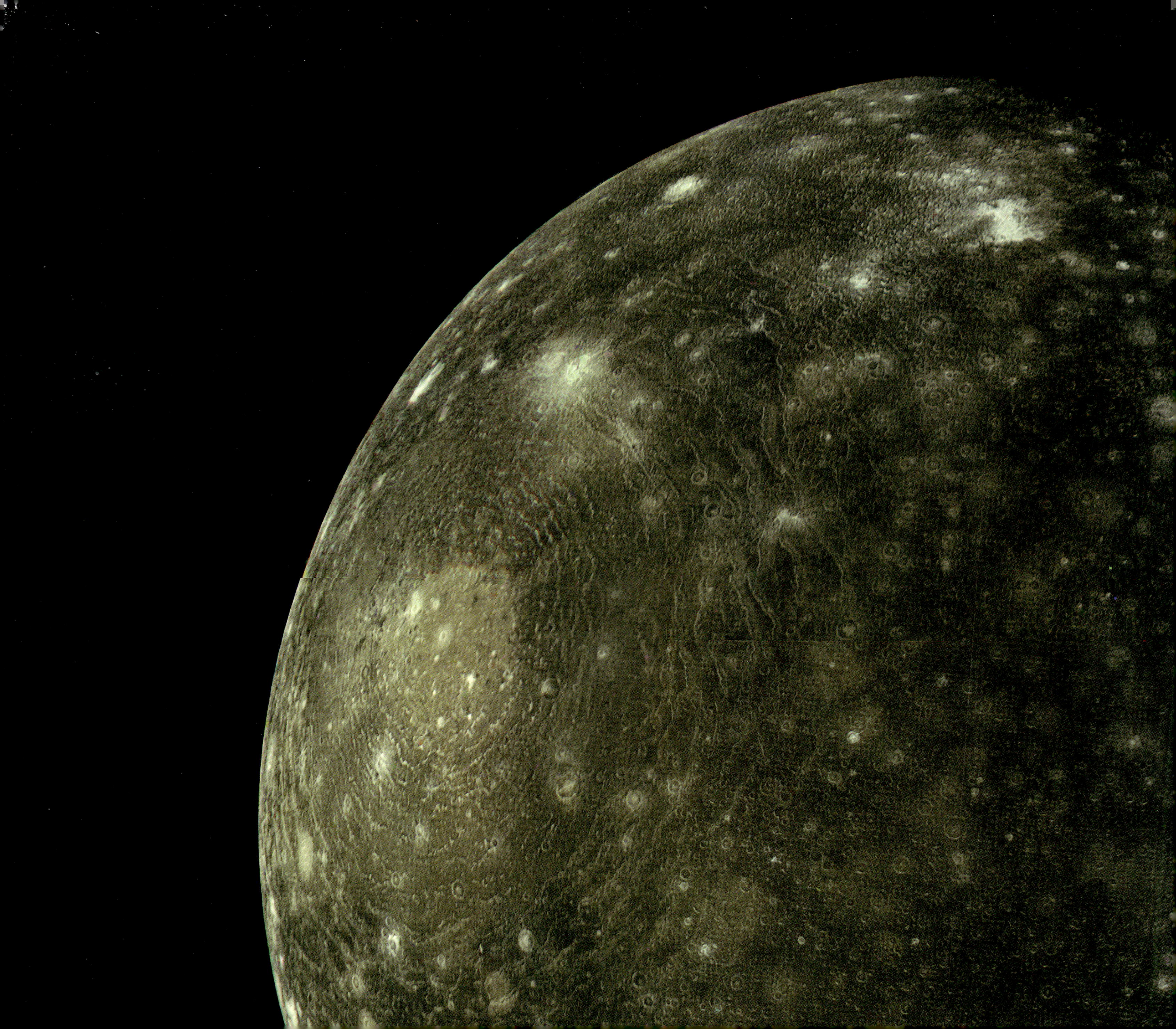

SATURN – LORD OF THE RINGS

Saturn is the second biggest and arguably the most beautiful planet in our solar system. Although its interior is spacious enough to accommodate 840 Earths, its mean density is only 0.7 grams per cubic centimeter. If we could place it in a tub of water, it would float. The reason: the mass of Saturn is only about one third that of Jupiter, which means that its interior is less dense because it is subjected to lower pressures. Like Jupiter's, Saturn's atmosphere consists almost entirely of hydrogen with some traces of helium. Saturn too has a core of silicate rock and ice which, however, is smaller than that of Jupiter. Violent storms rage through the atmosphere, distending the clouds into wide bands. However, Saturn's clouds are not as colorful as those of Jupiter. They mostly consist of ammonia congealed into crystals at temperatures around minus 190 degrees centigrade.

Even a small telescope will show the rings that surround Saturn. In fact, there are thousands of separate rings in which particles ranging in size from dust grains to family homes orbit the planet like miniature moons. The ringed planet has 61 known moons.

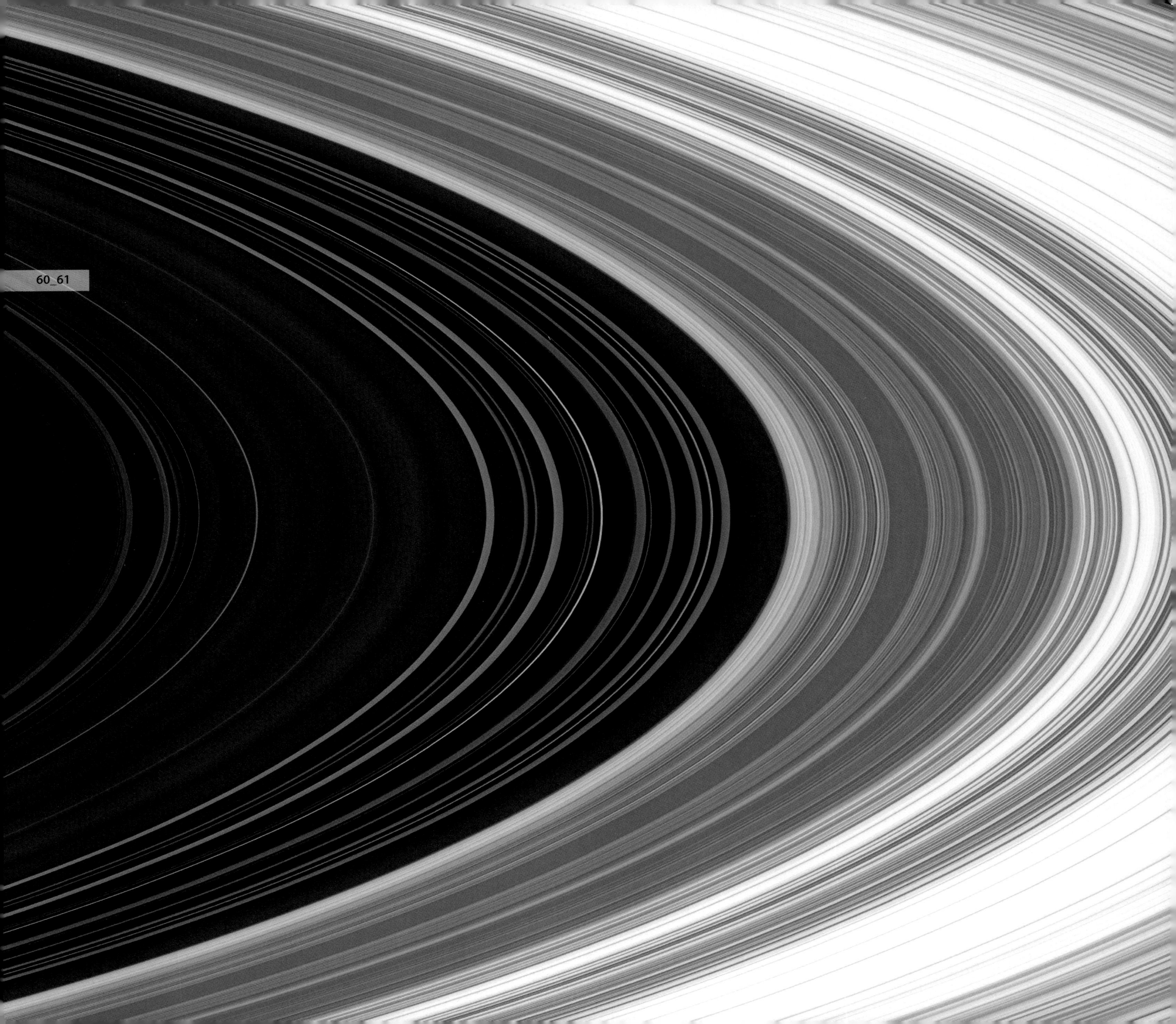

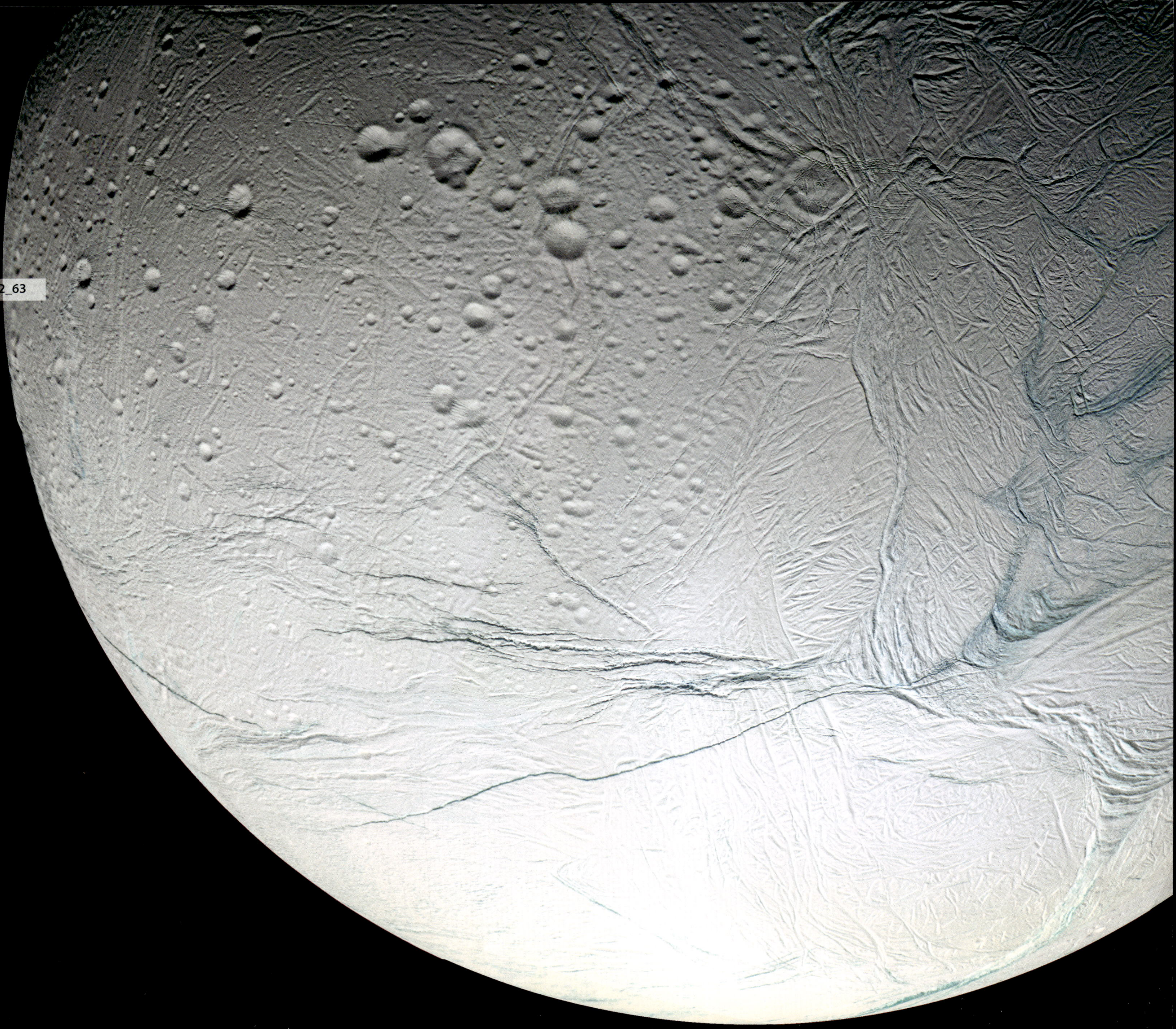

URANUS – THE BLUE SPHERE

Uranus appears as a globe of solid blue, although its color is not due to the presence of water. Its extensive atmosphere consists mainly of hydrogen and helium. At temperatures as low as minus 215 degrees centigrade, methane freezes out, forming haze particles that give the planet its blue color.

Scientists think that the gaseous atmosphere may surround a global ocean of hot water, molten rock and ice. At the bottom, this ‚sea' may conceivably hide a rocky core about four times as heavy as Earth.

Surprisingly enough, the planet's spin axis is inclined 98 degrees against its orbital plane. Rolling along like a cartwheel, the planet rotates backwards against the direction of its orbital movement. No conclusive explanation for this phenomenon has been found so far. It may be that the planet acquired a ‚list' when it collided with a large heavenly body. This inclination is the reason why, during the planet's 84-year orbit, its poles face towards the Sun about half the time and away from it for the other half, so that a day at the poles lasts around 40 years.

Like the other gas planets, Uranus is surrounded by rings, and it is known to have 27 moons.

NEPTUNE – THE TWIN OF URANUS

About as big as Uranus, Neptune has the same characteristic blue color. However, fierce storms are blowing on Neptune. In its clear atmosphere, white clouds of methane ice race around at up to 600 km/h. In 1989, the Voyager 2 space probe discovered a dark spot as large as Earth which recalled Jupiter's great red spot. This funnel-shaped cyclone was surrounded by white cirrus clouds of methane ice that are also known as scooters. However, the great dark spot was not as long-lived as its counterpart on Jupiter; it dissolved somewhat more than ten years ago.

Neptune's interior resembles that of its neighboring planet, Uranus. Moreover, Neptune is also surrounded by a few thin rings, and one of its 13 known moons, Triton, is the largest in the outer solar system.

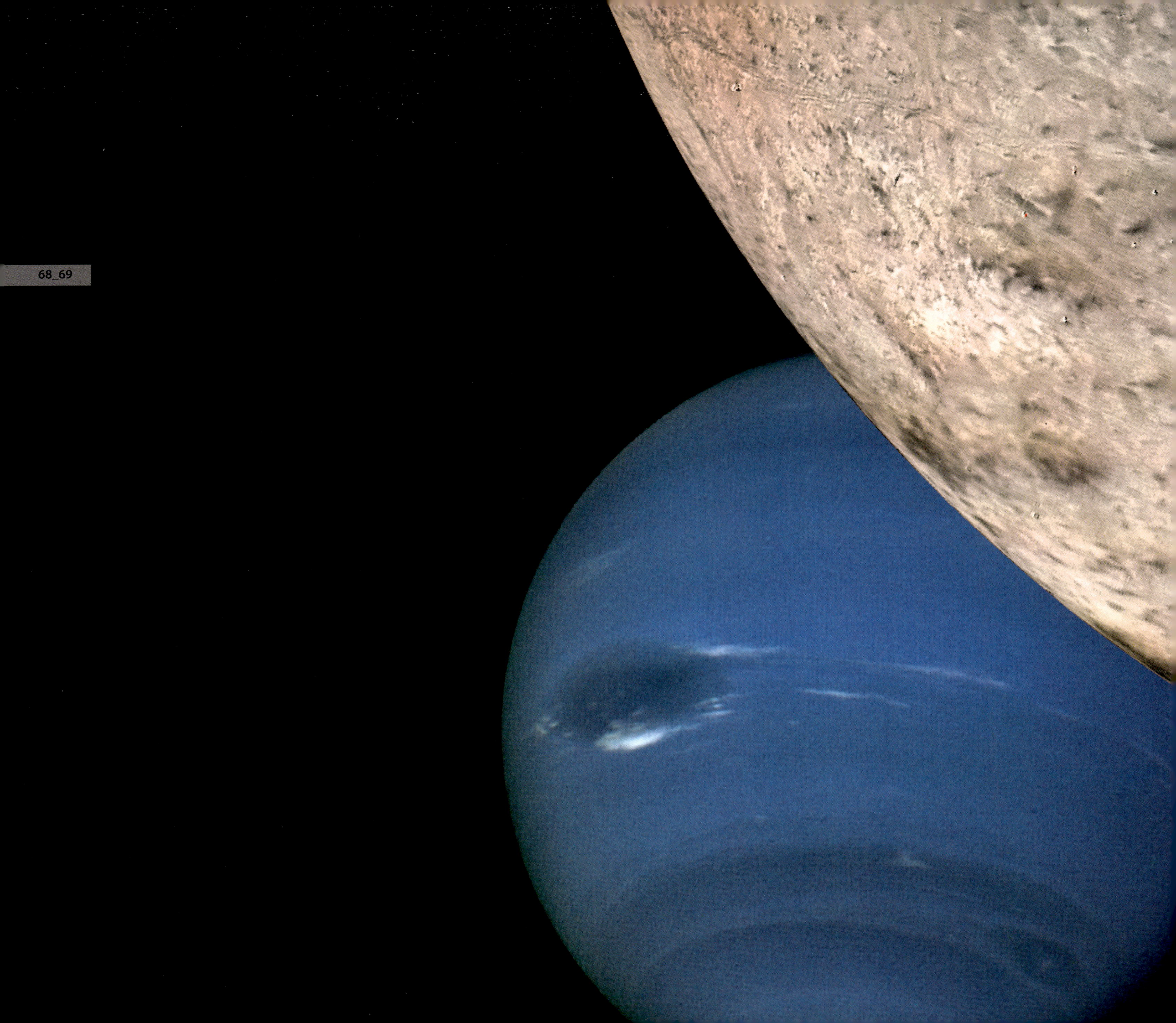

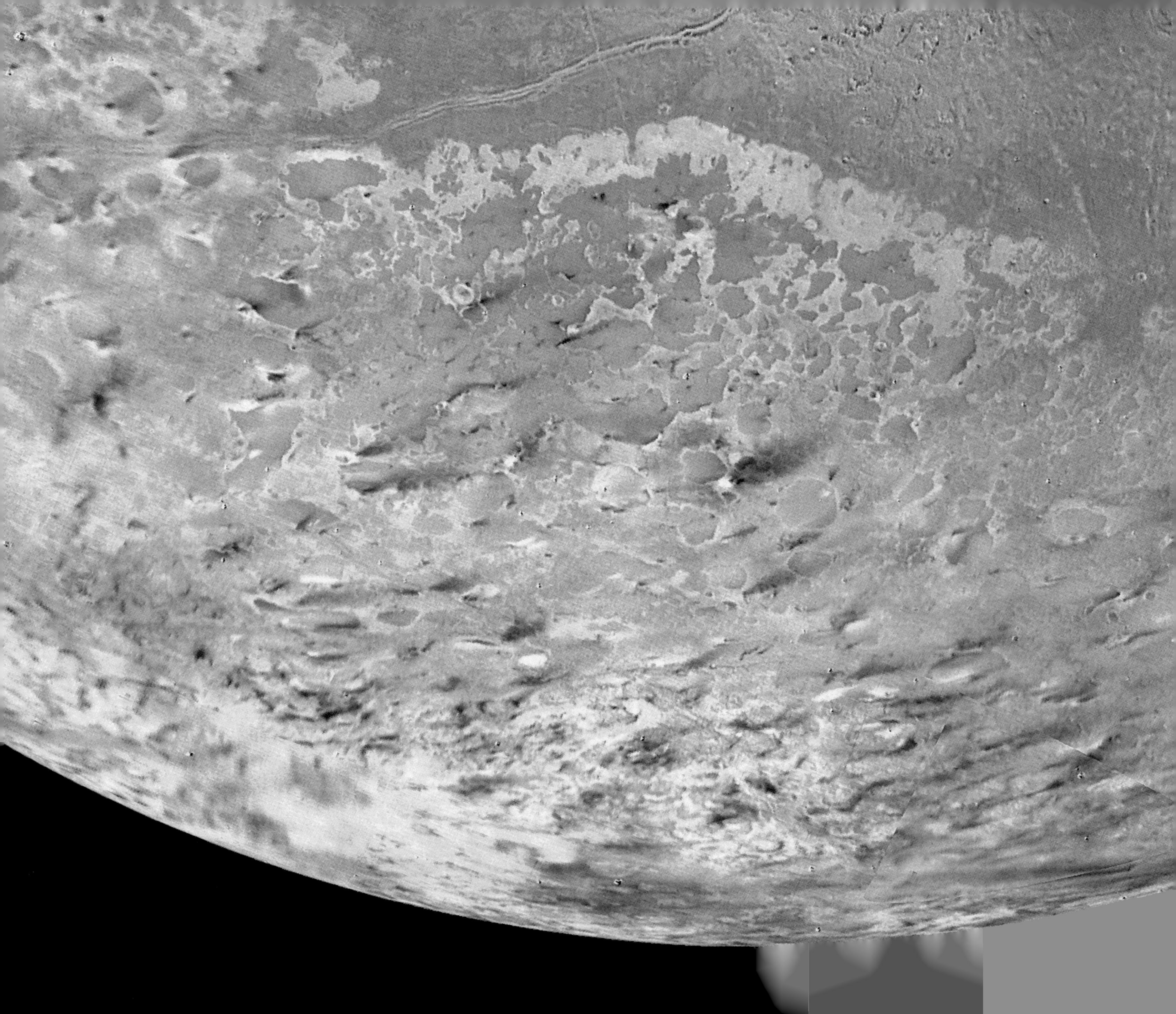

COMETS AND ASTEROIDS – THE MINOR MEMBERS OF THE SOLAR FAMILY

Comet bodies are generally no larger than a few kilometers. As they consist of rock, dust and ice, astronomers like to call them ‚dirty snowballs'. When a comet approaches the Sun, its temperature rises. The ice turns into gas, expands and shoots out of the core like a geyser. Comets are highly interesting objects for scientists because they have remained largely unchanged since their formation. They are something like natural deep-freeze archives. When astronomers demonstrated the presence of organic molecules in comet tails, the hypothesis that life on Earth might have arrived from space acquired a new lease on life.

More than one million asteroids measuring more than one kilometer in diameter travel around the Sun between the orbits of Mars and Jupiter. Scientists think that Jupiter's powerful gravity might have prevented all these small bodies from merging into a large planet in its vicinity, so that the asteroid belt formed instead. However, some asteroids travel on highly elliptical paths that cross the orbits of the inner planets, Earth included. Consequently, they may collide with our planet. If they do, small pebble-sized meteoroids flash up briefly as shooting stars. When, however, an asteroid measuring one kilometer or more in diameter crashes on dry land or into the sea, its inconceivably powerful impact triggers global consequences.

72_73

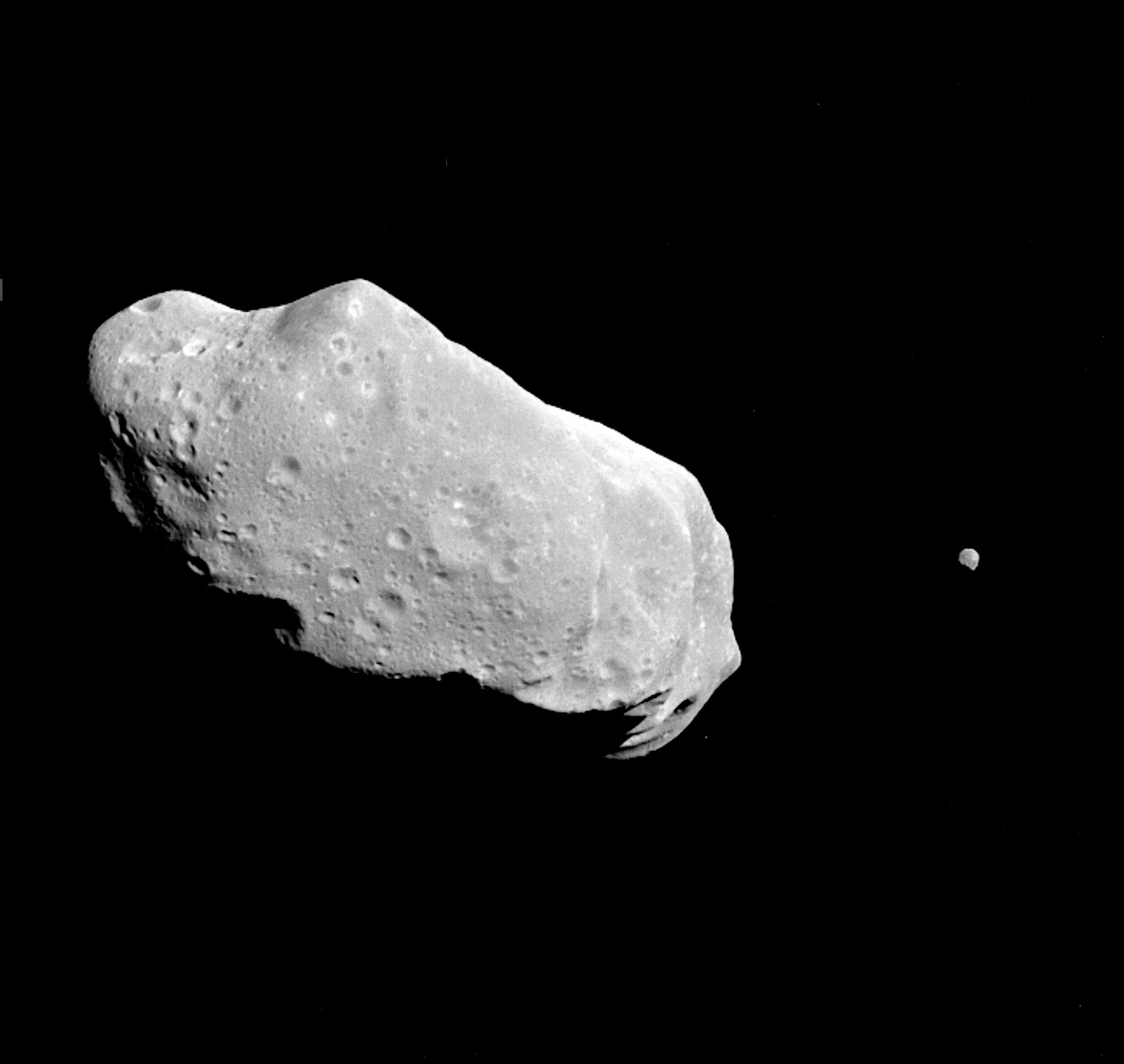

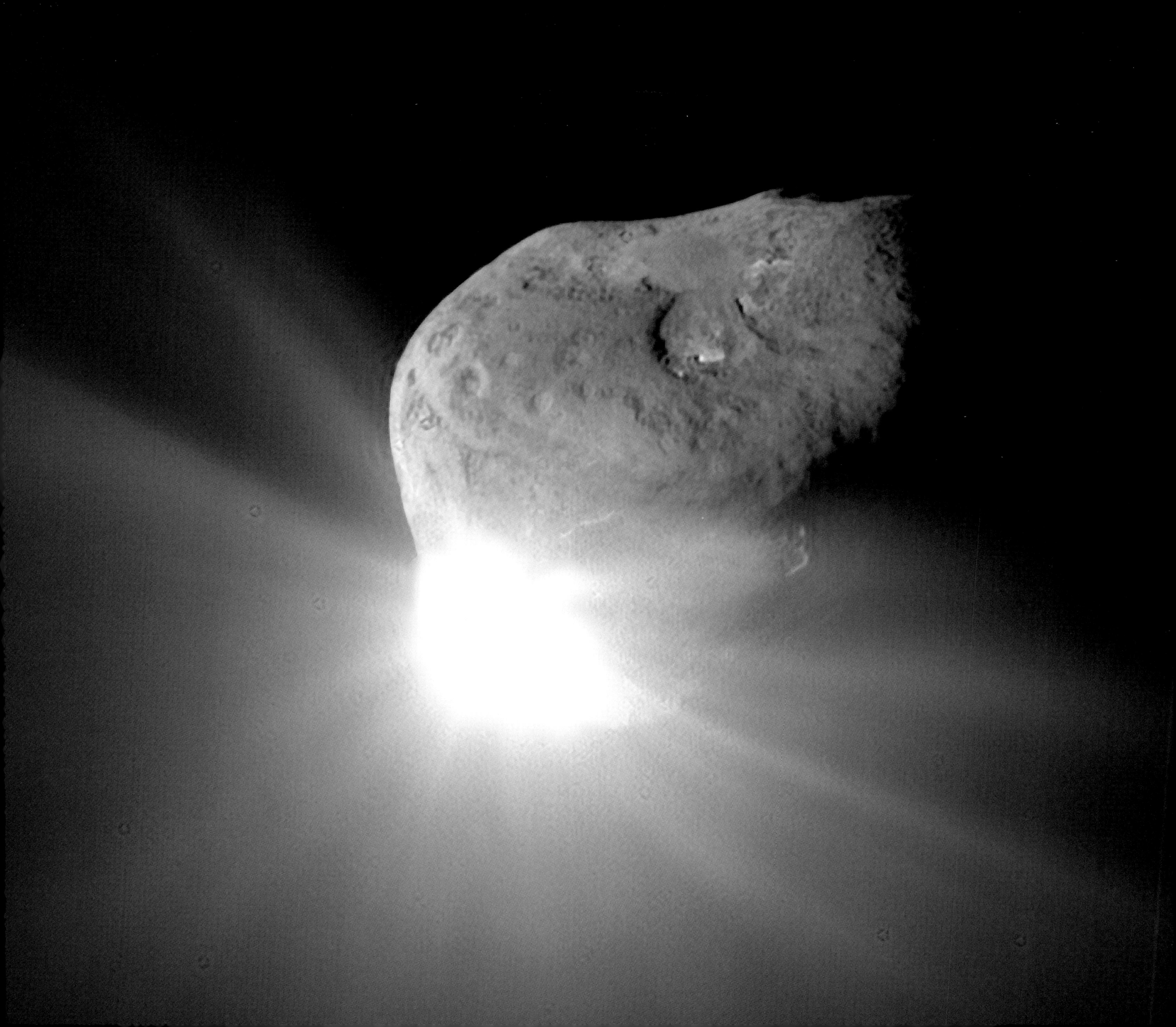

THE EARTH – OUR HOME IN THE COSMOS

Seen from space, the Earth appears as a blue planet. This impression is mainly due to the fact that nearly three fourths of its surface is covered by water. Moreover, the scattering effect of the atmosphere on the blue fraction of sunlight is particularly marked, which is why a cloudless sky appears blue to us when we view it from Earth. As far as we know today, our planet is the only one in the entire solar system to harbor life, mainly because of its gentle climate. The average temperature worldwide is 15 degrees centigrade, so that on most of our planet our elixir of life is available: liquid water. For this agreeable climate, we have to thank the greenhouse effect which is caused mainly by water vapor, carbon dioxide and methane contained in the atmosphere. Without this natural atmospheric greenhouse effect, the global average temperature would be minus 15 degrees centigrade, and the Earth would be a ball of ice.

The Earth's surface is subject to continuous change. Landscapes are molded by wind and weather, vegetation creates biotopes and plate tectonics pushes up mountain ranges and causes volcanos to erupt. There are two sources providing the requisite heat in the interior of our planet: residual heat from the formative period and decaying radioactive elements.

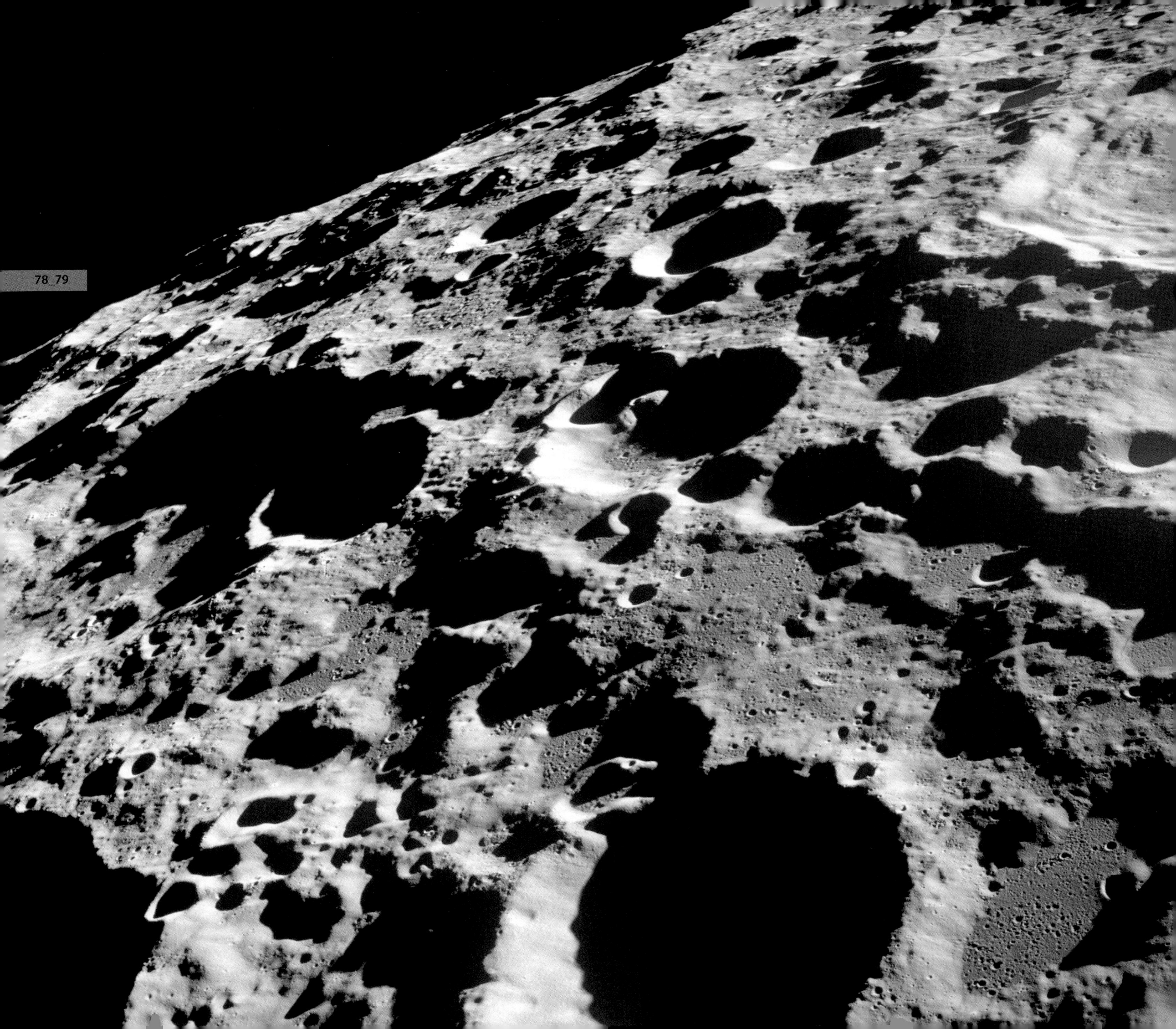

84_85

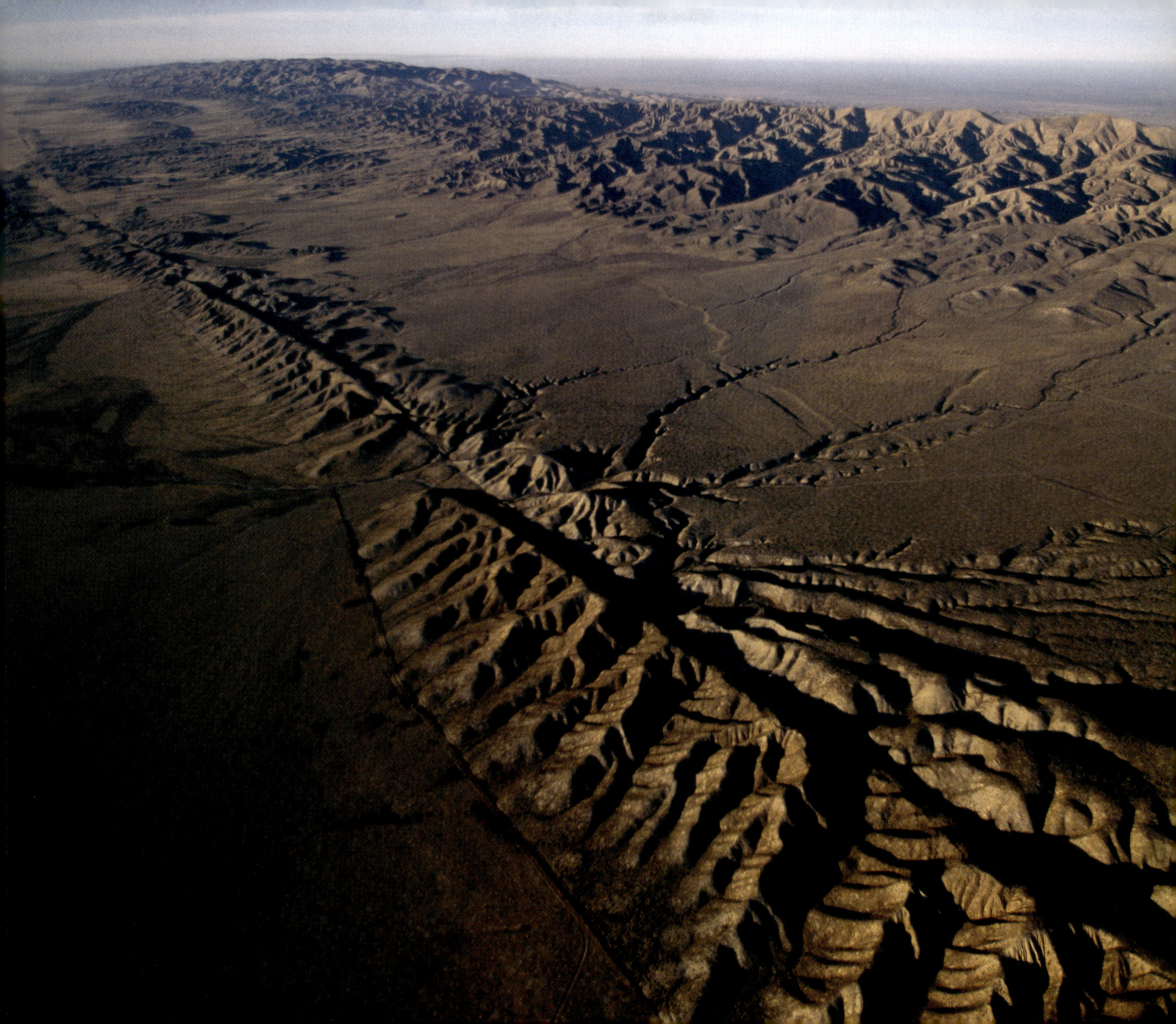

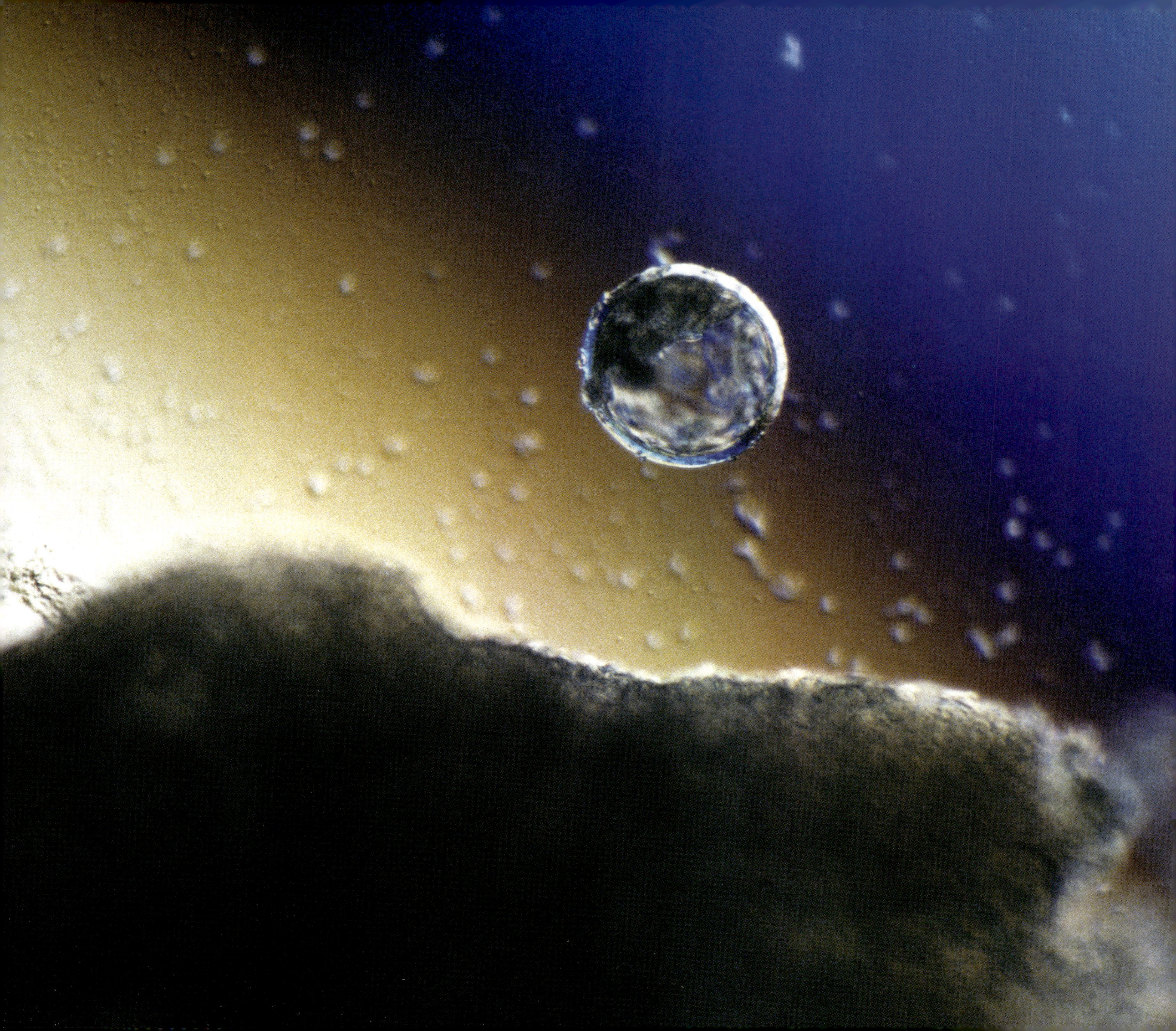

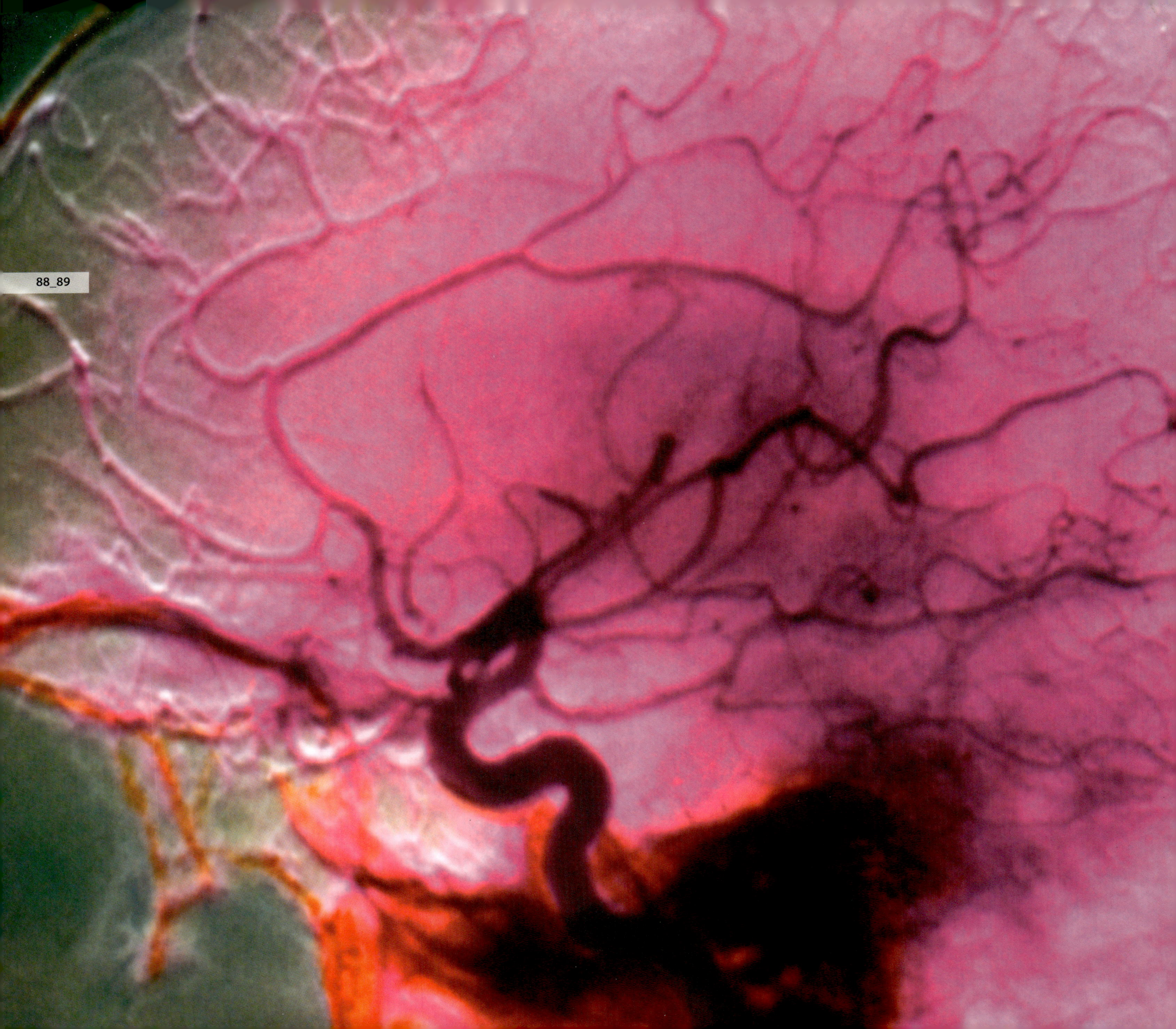

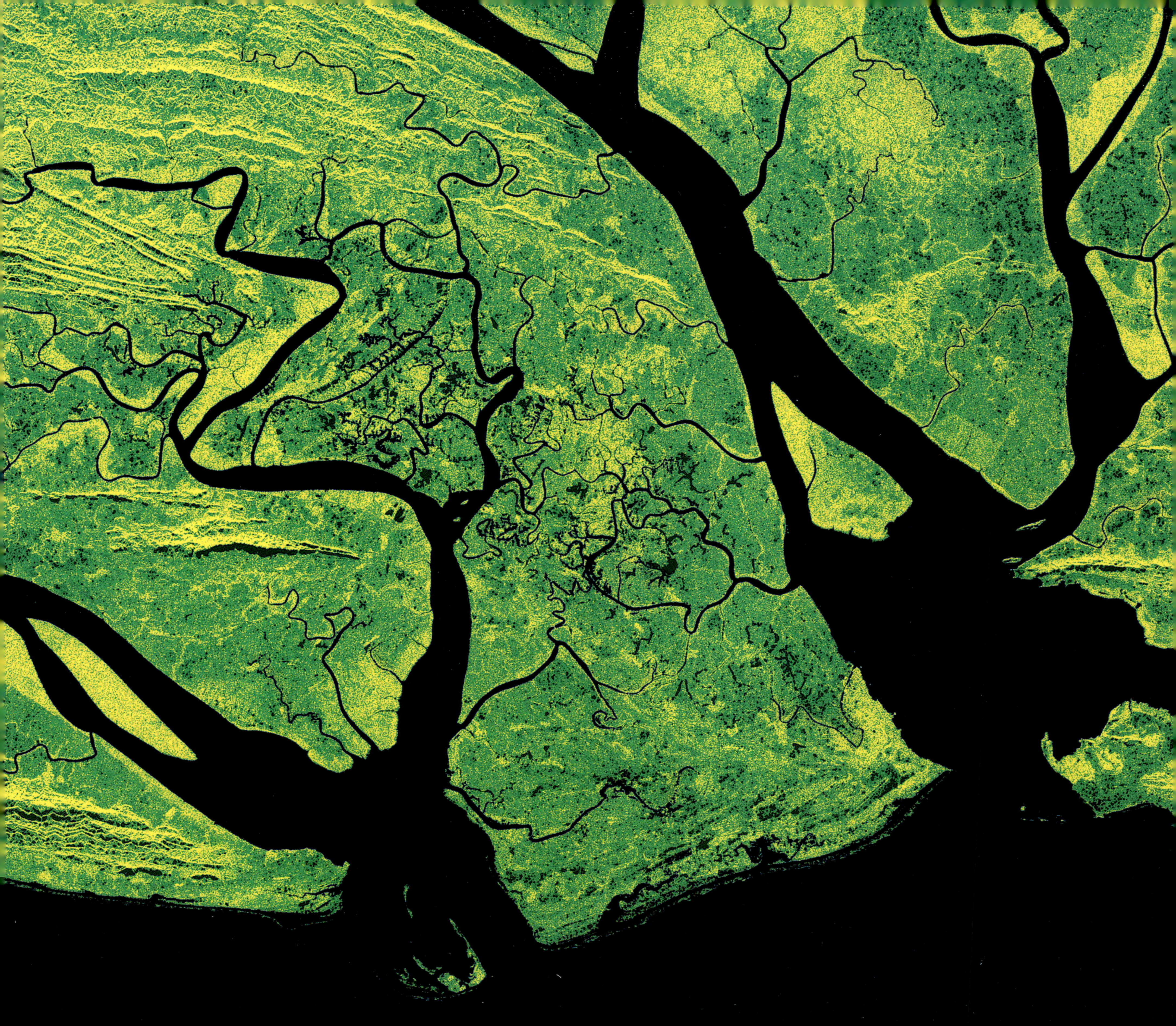

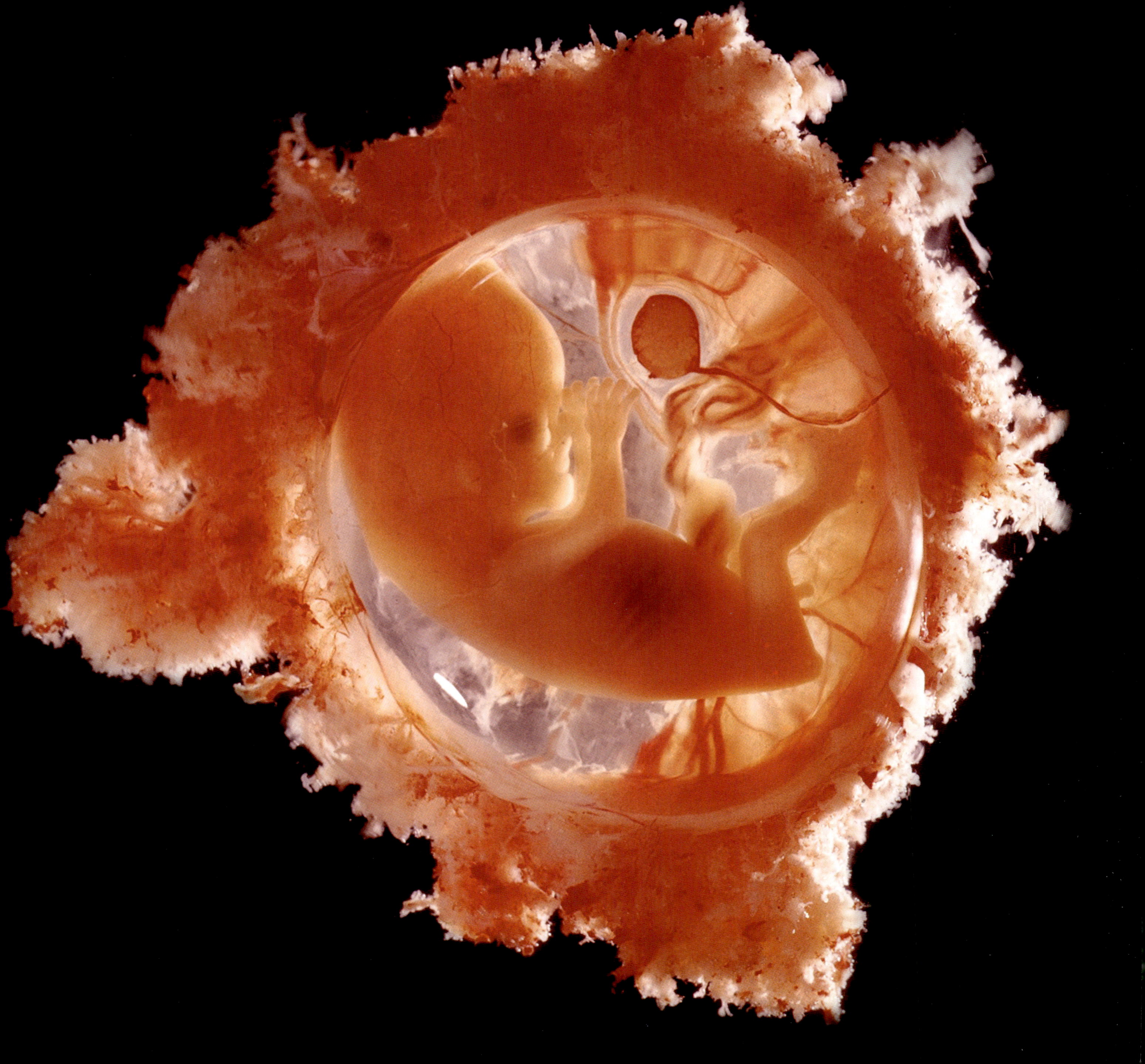

STELLAR DEATHS AND REBIRTHS

The matter contained in the universe is locked in a cycle whose dimensions are incomparably greater than anything we can imagine. The Big Bang produced nothing more than the volatile elements hydrogen and helium together with infinitesimal quantities of beryllium and lithium. There were, however, none of the heavier elements like nitrogen, oxygen, silicon or iron that are indispensable for the development of solid bodies such as planets. The first stars in the universe produced these substances, which they then emitted through either particle winds or supernova explosions. From then on, these newly-generated materials were available for the next generation of stars. In this way, each generation continued to enrich the universe with heavy elements until it was possible for planets to form. This ‚cosmic recycling process' is active even today, supported especially by planetary nebulas and the clouds of supernova explosions. At the same time, part of the matter thus dispersed is gradually withdrawn from the cycle by, for example, cooling white dwarfs or planets.

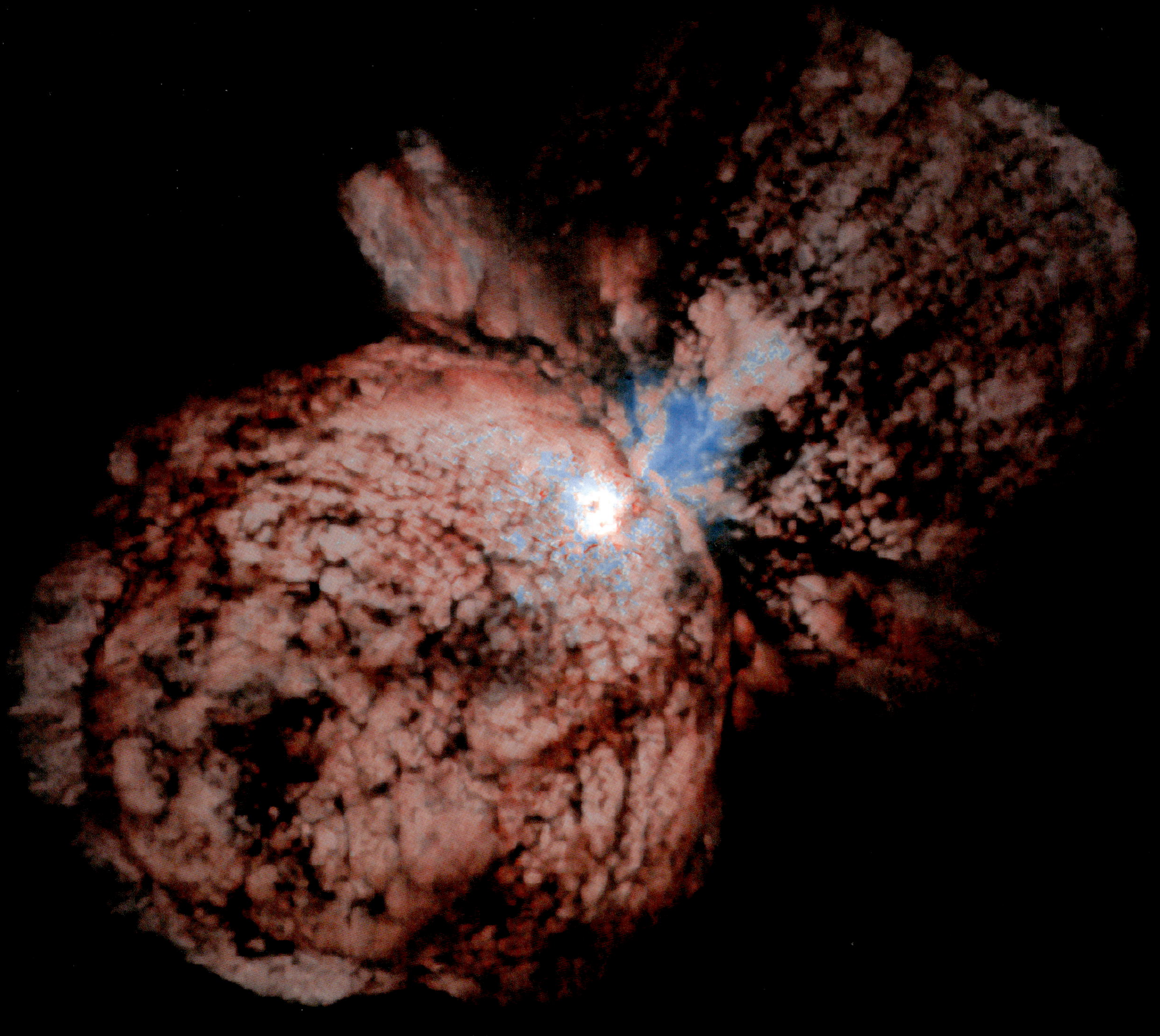

100_101

102_103

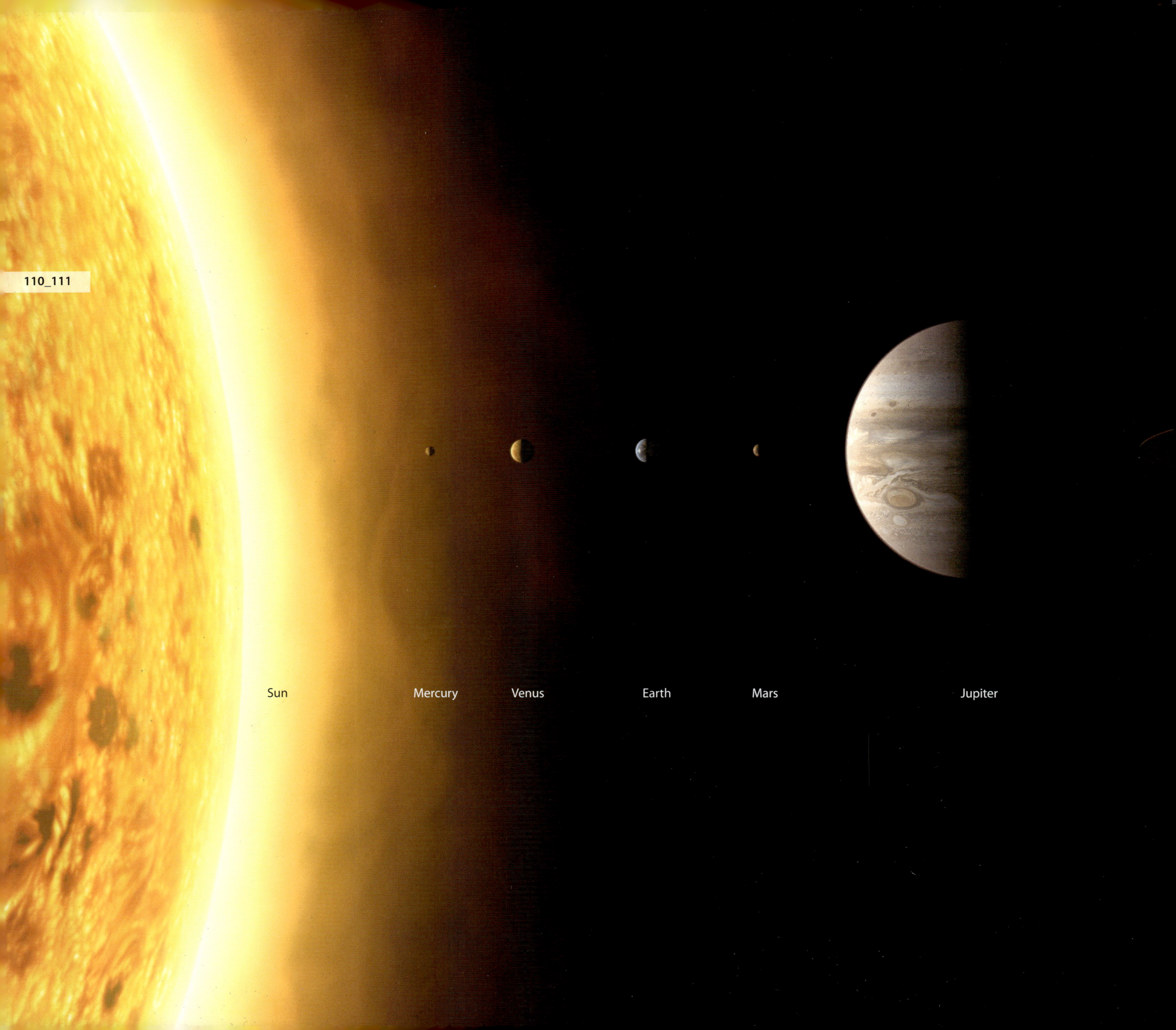
110_111
Sun
Mercury
Venus
Earth
Mars
Jupiter

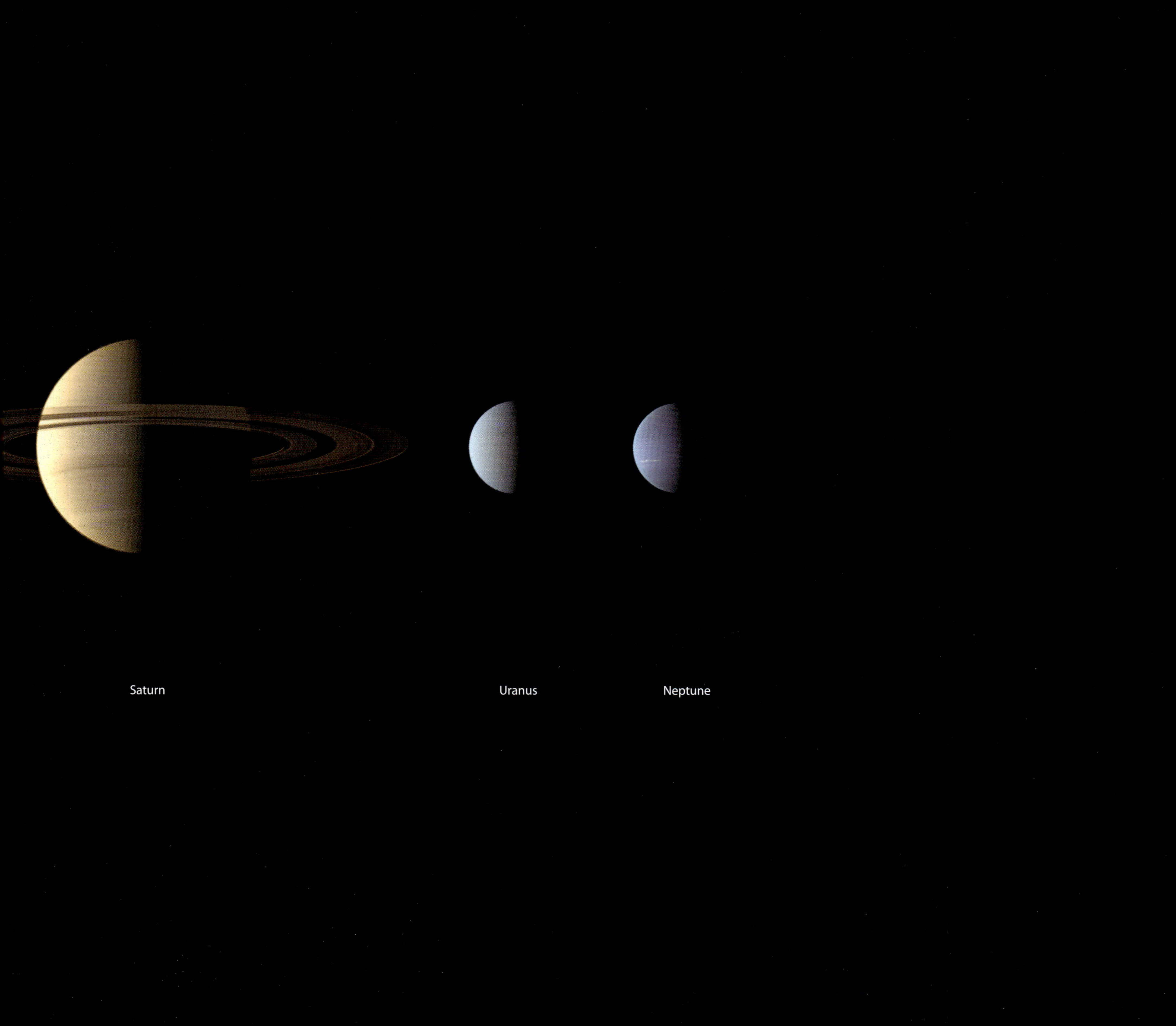
Saturn
Uranus
Neptune

PERSPECTIVES

The Oberhausen Gasometer is one of Europe's most spectacular exhibition sites. The gas-holder building, originally erected in 1929, is 117 metres tall and measures 68 metres in diameter. Listed as an industrial monument, the building was converted into a modern centre for art and history exhibitions in 1994. It has a total floorspace of 7 000 m^2 on three levels, with a 90 metre air space towering above the exhibition zone.

A walk through the exhibition „Out of this World – Wonders of the Solar System" starts at the base of the Gasometer and takes visitors along an extensive, sprawling installation of our Solar System and an array of large-format photographic tableaus illustrating the way in which our Sun and its planets developed. Pictures, objects and instruments displayed on the middle level document how man's concept of the Solar System has evolved over time: from the tradition of ancient civilizations to the creation of myths of the world religions to the body of knowledge amassed in the age of science. Scientific instruments used for the observation and measurement, past and present, demonstrate that human knowledge about cosmic processes has increased in leaps and bounds with the advent of every new observation and measurement technique, each time revolutionising our view into the depths of the Universe.

Wolfgang Volz's monumental installation of the Moon turns the airspace above the „circus ring" into a cosmic adventure. Visitors may choose to walk round the Moon sculpture while enjoying a composition of light and sound effects, go right through it in a glass elevator, or simply lie on their back on the floor in the ring, gazing up at it in admiration.

The illustrations on pp.113–120 show the Moon installation by Wolfgang Volz, a heliometer (Argelander Institute for Astronomy, Bonn), an astrolabe (Museumslandschaft Hessen Kassel), a celestial globe (Argelander Institute for Astronomy, Bonn), a tellurium (Wuppertal Clock Museum, an armillary sphere (CHRONUS-Manufaktur), the ExoMars Rover (German Aerospace Center, DLR).

114_115

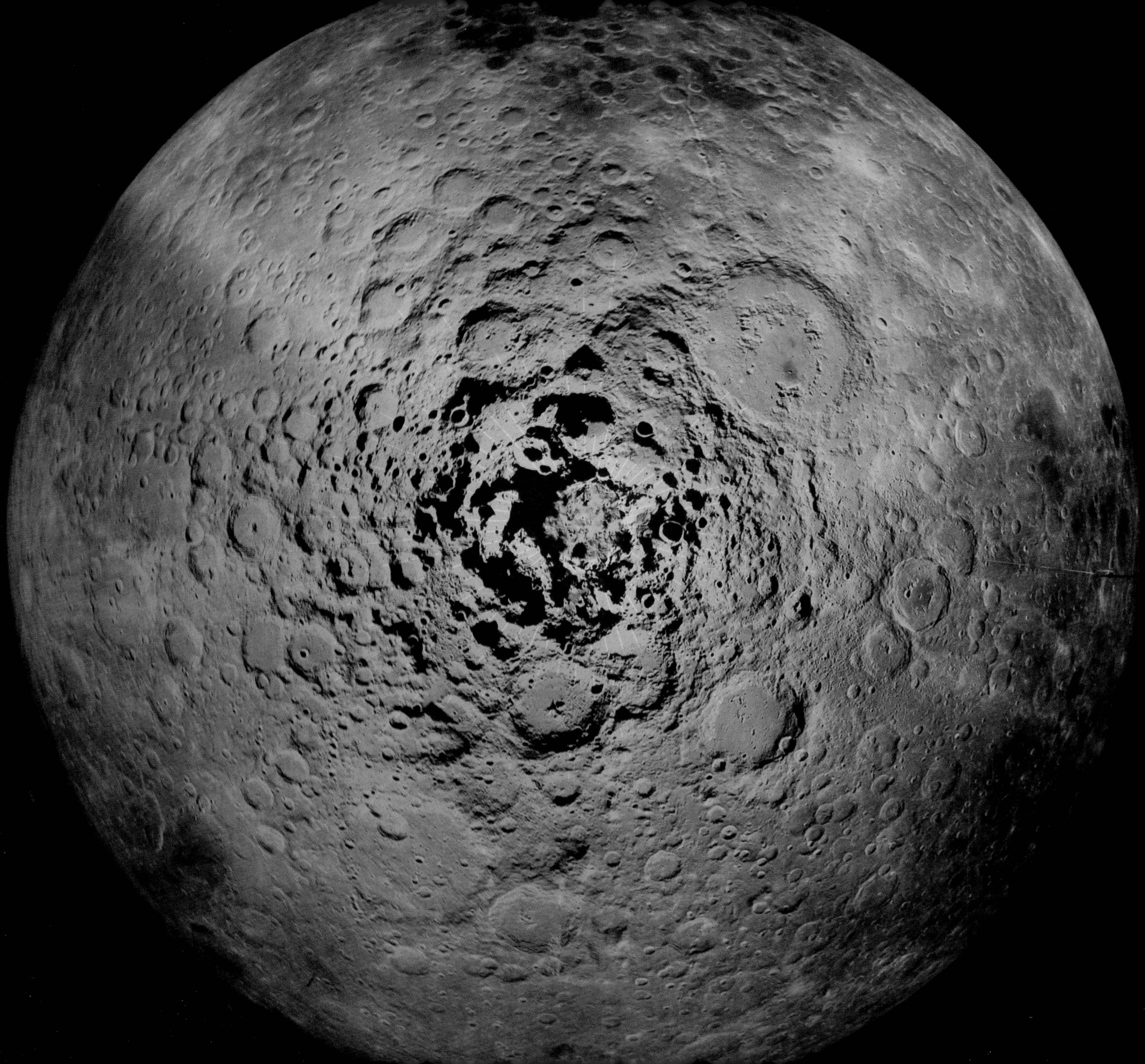

THE LARGEST MOON ON EARTH

The arena of the Oberhausen Gasometer with its 90-meter-high ceiling has always been a place where spectacular artistic displays were staged. This enormous void, used until 1988 to store nearly 350,000 cubic meters of coke oven gas, forms the highlight of any exhibition. When „The Eye in the Sky – Satellite Images of Earth" was shown in 2007/2008, light artist Christina Kubisch turned the Gasometer into a cosmic pantheon with her „light sky" In „Out of this World – Wonders of the Solar System", Wolfgang Volz' monumental sculpture of the moon transmutes the dome of the Gasometer into a planetary experience.

A photographer and close associate of Christo and Jeanne-Claude, Wolfgang Volz knows everything there is to know about realizing large-scale artistic projects. Moreover, several exhibitions have earned him his „Gasometer spurs". For years, he had been dreaming of realizing a sculpture of the Moon in this unique exhibition space. Close and yet mysteriously remote, the Moon fascinated him and always played an important part in his photographs. Now his dream has been realized: measuring 25 meters in diameter, the „largest moon on Earth" is hanging suspended inside the Gasometer. Its lighting shows visitors the phases of the moon in all their splendid color, while Ulrich Sende's music enhances the sensuous experience of the sculpture. Dominating the airspace within the Gasometer, the Moon emphasizes its incredible dimensions.

The production of the Moon sculpture was a protracted affair. The Moon data provided by the German Aerospace Center (DLR) had to be processed and converted painstakingly and carefully before they could be transferred to the skin of the balloon, whose material composition had been determined previously in numerous tests conducted by balloon engineers.

On February 16, 2009, the constituent parts of the sculpture were brought to Oberhausen, where the Moon rose in the Gasometer within two days. In a last step, the lighting technicians had to do some fine-tuning to ensure a seamless cycle from the full to the new moon.

Wolfgang Volz' dream corresponds with another dream about the Moon: it was 40 years ago, on July 21, 1969, that man first set foot on the Moon. So there is more than one reason for „capturing" the Moon in the year of astronomy. „Out of this World – Wonders of the Solar System" was born from the exhibition „The Eye in the Sky".

In „The Eye in the Sky", satellite images of rousing beauty afforded fascinating views of our Earth. Positive feedback from 375,000 visitors was the reason why we began considering a follow-up exhibition together with DLR. This was the time when the idea of changing our perspective was born. If it was so fascinating to view our home planet from a heavenly perspective, it should not be any less interesting to look into space, at our solar system and our neighboring planets. Initial analyses convincingly proved that pictures showing the Milky Way, the birth of stars, the Sun, the planets and their moons were of breathtaking beauty.

The two curators, Peter Pachnicke and Wolfgang Volz, developed the concept of the exhibition from this idea, which was realized in cooperation with DLR scientists and the members of the scientific and cultural advisory board with the support of major loan providers.

When Europe's tallest disk-type Gasometer was commissioned in 1929, it was a hallmark of Oberhausen's industrial development. In 1988 its closure marked the end of the industrial era. At the time, nobody could have imagined that the Gasometer would have another, even more exciting life as an exhibition hall. Today, after 3.2 million visitors and 10 exhibitions, it is once again a brightly shining star in the sky above the Ruhr basin.

Visitors of the „Out of this World" exhibition may now abandon themselves to the Moon, looking at it from the arena, orbiting it on the elevated walkway, traveling up and down in a glass lift cage, and admiring it in wonder from a supine position on the floor of the arena.

The Moon sculpture and the fascinating pictures have brought the cosmos to this cathedral of industry. What visitors experience is „Out of this World – Wonders of the Solar System".

JEANETTE SCHMITZ

GROWTH AND DECAY IN THE COSMOS

The song „Woodstock" by the pop group Crosby, Stills, Nash and Young contains this verse:

We are stardust, we are golden.
We are billion year old carbon.

These two lines express in poetic terms one of the most astonishing insights of modern astrophysics. Each and every carbon atom in our bodies originated billions of years ago in the interior of some star. Later, the explosion of the star or some other mechanism ejected them into space, where they gathered in an enormous cloud together with many other atoms. When the cloud grew beyond a certain size, it contracted under the influence of its own gravity until a new star was born: our Sun. It was surrounded by a rotating disk of gas and dust in which lumps formed that later evolved into planets, moons, asteroids and comets.

The stars „bred" not only carbon in their interior but also all the other heavy elements like oxygen, nitrogen and iron. The Big Bang, the hot beginning of the universe, produced only hydrogen and helium as well as small quantities of lithium and beryllium. Solid heavenly bodies could never have formed from these alone.

In about six billion years' time, our own Sun will have consumed its fuel and expand into a red giant. At that time, it will discharge the outer part of its gaseous envelope into space, yielding up part of its substance as building material for a new generation of stars and planets and closing the cycle of growth and decay in the universe.

Persistent change also features in the development of planets and moons. Our own Earth is the best example: landscapes are molded by wind and weather, and vegetation creates biotopes. Plate tectonics, a phenomenon that exists only on Earth and nowhere else in our solar system, has left its mark on the evolution of life. Fitting together like a global puzzle, continental plates float on an elastic layer called the astenosphere. Wherever two plates drift apart, long rifts open up in which hot crust material rises, pushing continents apart by as much as 10 centimeters per year. In other places, one plate slides under another and dives into deeper layers. In such places, mountain ranges like the Himalayas rise up. By overturning rocky material, plate tectonics integrates it into a never-ending cycle.

Other heavenly bodies in our solar system are similarly subject to change. Thus, volcanic activity is greatest on Jupiter's moon Io. Its volcanoes eject enough gas and dust to cover the entire surface with a one-centimeter layer in the course of 10 years. Another spectacular heavenly body is Saturn's moon Titan. The rain on it probably consists of liquid methane which collects in gigantic lakes. The periodic growth and contraction of the polar ice caps on Mars and the dynamic processes in Jupiter's atmosphere are only two of many more such examples of change.

Located at the center of our planetary system, the Sun plays the leading role. The Sun is our source of light and heat, and we would not be around without it. Until a few decades ago, visible light was the only medium available to scientists for studying the Sun, and all they could see was its surface. Being a gas ball, the Sun does not have a solid surface in the conventional meaning of the term. Astronomers use it to designate the altitude range within which visible light is generated. Called photosphere, this layer is several hundred kilometers thick and has a temperature of 5,500 degrees centigrade.

Although it appears white, the light of the Sun is really a blend of many colors, as we can see if we look at a rainbow. This phenomenon can be explained by the fact that light consists of electromagnetic waves. Each color is characterized by the length of the respective electromagnetic wave. Red waves are longer than those of blue or violet light. This electromagnetic spectrum is in no way restricted to visible light. Thus, x-ray and UV radiation differ from visible light only in that their wavelengths are much shorter, while those of radio and infrared radiation are longer.

Each wavelength or spectral band opens its own window for observing space. As a rule of thumb, we might say that the shorter the wavelength, the higher the energy of the radiation and the hotter the body from which it emanates. In visible light, for example, we can see the Sun's 5,500-degree surface, the UV band shows us the chromosphere where the temperature reaches multiples of 10,000 degrees, and x-rays show us the gas in the corona where it is up to millions of degrees. This is why the astronomers of today can draw a 'multichrome' picture of the Sun. And this picture shows a heavenly body that is much more explosive and active than it appears to the naked eye.

Observation at different wavelengths has also opened up entirely new approaches in planetary research. Atoms and molecules absorb and emit light at highly specific wavelengths. These spectra are like 'fingerprints'. This being so, researchers can now identify the substances and minerals in the surface rock of a planet from these spectra.

Astronomers employ the same spectroscopic method to demonstrate the presence of chemical elements or molecules in remote stars or gas nebulas. Scientists use it to study the process by which stars eject enormous amounts of heavy elements like carbon into space through particle winds or explosions. At some other time and place, these may be involved in the development of a new lifeform which, pondering its origins as we do, may see that

We are stardust, we are golden.
We are billion year old carbon.

DR. THOMAS BÜHRKE

The four Galilean moons – the four largest moons were discovered by Galileo Galilei exactly 400 years ago:
Callisto, Ganymede, Io and Europa (from left). Together with their central planet, Jupiter, they form a kind of miniature solar system.
(Photo: Galileo space probe, NASA, JPL, DLR)

2_3

Our cosmic home

Forming an almost flawless arc, the Milky Way spans the starry sky. It is our cosmic home: a spiral galaxy measuring about 100,000 light years in diameter which contains some 100 billion stars. One of these is our Sun. The Milky Way's shape resembles that of a discus. If we look up along its central plane, we see a great many stars, but only a few if we look vertically away from it. The dark patches in the band of the Milky Way are dust clouds where new stars and planetary systems are being born. The center of the Milky Way is visible just above the horizon on the right. To the left, the planet Jupiter shines brightly, while Antares can be seen on the right. Although they appear close together in the sky, their distance from us differs greatly: Jupiter is about 40 light minutes away, while Antares is 600 light years and the center of the Milky Way 28,000 light years distant. This panoramic view extends across some 250 degrees along the horizon. The mountain on the left is the 5,671-meter Demawend in the Elbrus range (Iran). Towns light up the landscape here and there.

Photo:
B. Tafreshi, TWAN

5

Venus in transit

Passages of the planet Venus before the Sun, also known as Venus transits, happen only three times every 130 years. Taken on 8 June 2004, this photo illustrates the relative size of the two heavenly bodies, besides showing some activity and movement on the Sun itself. At its edge, an atmospheric layer about 10,000 kilometers thick appears as a reddish fringe: the chromosphere. In it, clouds of hot gas that are variously called spicules or protuberances rise explosively. While they are most clearly visible at the edge of the Sun, they occur all over its surface, making it appear like a wheat field waving in the wind. This photo was taken through a special filter which permits only the light of hot hydrogen to pass.

Photo:
S. Seip – astromeeting.de

13

Stars being born in Orion

New stars are evolving everywhere in the Milky Way. Astronomers estimate that the birthrate ranges around five stars per year. One of the largest star cradles is located in the Orion constellation. The clouds that glow red consist of hydrogen, the most prevalent element in the cosmos and the raw material for its stars. The black silhouette that rises at the lower edge of the nebula on the left is the famous horsehead nebula, a dense cloud of dust three light years in diameter. The Orion nebula, which is 1,500 light years distant, appears close to the right edge of the photo. At the center of this bowl-shaped cloud of gas and dust, four young stars shine in a configuration called the trapeze. All in all, several thousand young stars populate the interior of this nebula which is 30 light years in size.

Photo:
R. Gendler

15

The elephant's trunk

Extending across some 20 light years, this cloud of gas and dust is located inside a giant bubble of red-glowing hydrogen which is heated and induced to glow by young hot stars. Called the elephant's trunk because of its appearance, this longish cloud of dust is nevertheless very cold inside. Here, matter may contract to form new stars. Although the elephant's trunk appears like a dense cloud of smoke, its dust is spread out more finely than in a clean-room laboratory on Earth. It is only its enormous size that makes it appear almost opaque. Only the brightest stars in the background can penetrate it, with their light taking on an orange-yellow hue like that of a sunset.

Photo:
Canada-French-Hawaii-Telescope,
J.-C. Cuillandre, G. Anselmi

16_17

The stuff stars are made of

About 7,600 light years distant, this large nebula in the Carina (keel) constellation is one of the most active among the star nurseries we know, where the cycle of growth and decay is operating. Dark dust clouds float before a background of radiant hydrogen gas. Some of these clouds show a bright fringe because they are heated and stimulated to emit light by hot stars in the vicinity. A heavenly body in which astronomers are most keenly interested shows at the center of the left edge: Eta Carinae. With a mass about a hundred times greater than that of the Sun, it is one of the heaviest and most luminous stars we know: in seven seconds, it emits as much energy as our Sun does in an entire year. In the past, Eta Carinae has erupted on several occasions, ejecting carbon-rich layers of gas into space which today surround it in the form of nebulas. Expanding further, this matter will serve as building material for future star generations. Having consumed most of its fuel in the past, Eta Carinae will explode in a supernova in the distant future, when it will be the brightest star shining in the night sky.

Photo:
Hubble Space Telescope, (STScl), NASA, ESA,
N. Smith (California U., Berkeley)

18_19

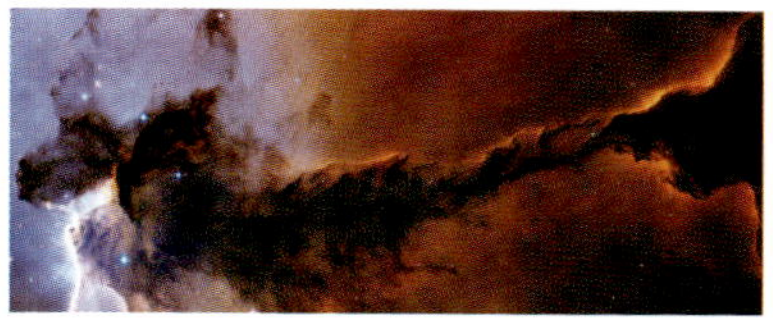

Pillars of creation

7,000 light years distant, the eagle nebula is composed of several longish pillars of dust. Some of these have been demonstrated to contain stars in the process of formation. For this reason, these clouds have been called pillars of creation. But the birth of these stars is a race against time. A star is born from dust and gas that contract under the influence of their own gravity. At the same time, the surface of the dust cloud is heated by stars that exist already, so that it begins to shine and evaporates gradually. Thus, adolescent stars are engaged in a race against the evaporation of their raw material.

Photo:
NASA, ESA, STScl, AURA

20

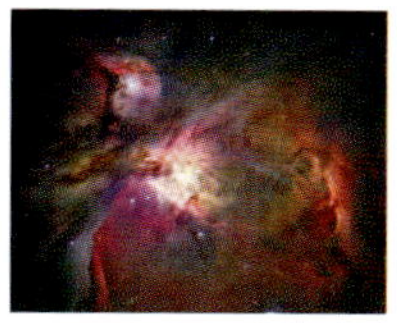

Budding planetary systems in the Orion nebula

In this image of the Orion nebula, the most highly defined so far, astronomers find a wealth of fascinating detail. The center is occupied by the four stars of the trapeze. Their intense radiation and the strong particle winds that emanate from them have created a bubble that surrounds them within the cloud of dust where they originated, so that we can see them clearly now. Another 3,000 or so young stars still remain hidden in the nebula. Once it has evaporated completed after some scores of millennia, this part of the sky will be lit by a bright star cluster. Moreover, it was the Orion nebula where astronomers discovered dust disks surrounding young stars for the first time. From these disks, planets will probably emerge. Clearly, these protoplanetary disks are now in the same stage of development as our own solar system about 4.6 billion years ago.

Photo:
Hubble Space Telescope, NASA, ESA,
M. Robberto, the Hubble Space Telescope Orion Treasury Project Team

21

A cosmic treasure chest

This impressive star cluster bearing the designation NGC 3603 looks like a hoard of sparkling jewels. Its center is occupied by several thousands of young stars. It is the largest agglomeration of extremely massive stars we know. Weighing up to a hundred times more than our Sun, these giants are very wasteful with their fuel. Whereas our Sun will reach an age of more than 10 billion years, giant stars expend their fuel within a few million years, after which they explode in a supernova, enriching their environment with chemical elements „bred" in their interior. From this material, new stars and planets will evolve.

Photo:
Hubble Space Telescope, NASA, ESA
STScl, AURA

23

The Sun's aureola

When the Moon, moving in front of the Sun, absorbs its dazzling light, an aureola appears: the corona. Visible only during a total eclipse, this halo of light has been fascinating mankind for millennia. Today, astronomers know that what we see is a kind of solar atmosphere consisting of highly attenuated gas at a temperature of several million degrees centigrade. Its radial structure is caused by the fact that its constituent gas particles (mostly hydrogen nuclei) follow magnetic field lines like iron filings above a bar magnet. At the edge of the Sun, moreover, arcs of gas (protuberances) appear that indicate explosive processes on its surface.

Photo:
S. Seip – astromeeting.de

24_25

Our explosive Sun

When we watch the Sun as it sets on the horizon it appears like a steadily glowing ball of gas. However, observations made at other wavelengths show that this is a false impression. Equipped to receive X-ray and UV radiation, the European-American space observatory SOHO generates images of the Sun in these wavebands. The picture above was taken on 14 September 1999 in the ultraviolet, which only shows hot gas at temperatures around 60,000 degrees. Areas that are particularly hot and explosive appear whitish. In many places, gas is flowing away from the surface, always guided by the lines of the Sun's magnetic field. Rising around 80,000 kilometers above the surface, the great protuberance shown here could accommodate nearly seven Earths under its arc.

Photo:
SOHO Space Telescope, ESA, NASA

26

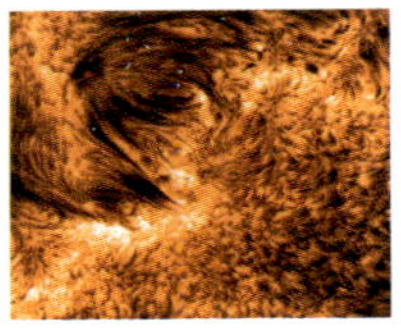

A boiling sea of flames

The Sun's surface is far from being as smooth as it appears to us. The dark arcs at the bottom right-hand edge of the photo are called spicules, tube-like structures rising as high as 10,000 kilometers above the surface in which hot gas shoots up like a fountain. Spicules exist only for about five minutes, but they are so frequent that they rear up like licking flames in their hundreds of thousands at any time. These jets of hot gas maintain a fragile equilibrium between the forces of gravity and magnetism. High up on the left we can identify the similarly arc-shaped structure of a sunspot. All these phenomena are caused by swiftly changing magnetic fields which attract gas particles. The area shown in the photo extends horizontally over 64,000 kilometers, a distance large enough to accommodate five Earths side by side.

Photo:
Swedish Solar Telescope La Palma, Royal Swedish Academy of Sciences, LMSAL)

27

Solar gas at millions of degrees

Seen in the light of X-rays, the Sun presents a startling picture. This high-energy radiation is emitted by gas at a temperature of about one million degrees, thus helping us to identify particularly active areas where radiation eruptions are going on. These appear as bright spots. The areas of activity are arranged along two specific degrees of latitude to the north and south of the equator. Caused by the magnetic field in the Sun's interior, this association with certain latitudes also applies to sunspots, most of which appear near centers of activity. At the edge of the Sun, the hot gas looks tattered, the reason being that most atoms in it are electrically charged, having lost their electron shells. In that state, they arrange themselves along magnetic field lines so that these become visible.

Photo:
Trace Space Telescope, Trace Project, NASA

28

Hot gas within magnetic field arcs

This UV image shows hot gas which, arranged along magnetic field lines, rises far above the surface of the Sun. The magnetic field is generated by the flow of enormous masses of gas in the interior of the Sun. Twisted together in the process, magnetic field lines occasionally rise to the surface from which they emerge as loops that form magnetic arcs. Where they emerge, sunspots form that appear dark because their temperature at the center is around 1,800 degrees centigrade lower than in the vicinity. The arc roots are not firmly anchored but move with the hot masses of gas that well up from the interior of the Sun. This being so, the Sun's magnetic fields jump about, twisting around one another like rubber bands. At some time or other, tension rises to a point at which field lines break up and reconnect again. This magnetic short-circuit causes part of the magnetic field to disappear, the energy stored in it is released at a stroke, and we have a solar eruption.

Photo:
Trace Space Telescope, Trace Project, NASA

29

Polar lights

The polar lights that appear in the high northern and southern latitudes belong among the nature's most spectacular phenomena. They are caused by the Sun. Explosions on its surface eject streams of particles into space at enormous speed. Reaching Earth within a few days, they can give its magnetic field a thorough shaking. At the poles, where the lines of the Earth's magnetic field bend down towards the surface, particles may enter the atmosphere where they collide with air molecules and induce them to glow. Green light is emitted by oxygen atoms, while nitrogen atoms shine in violet and blue. What is more, massive particle storms may interfere with the radio transmissions of aircraft and ships, cause power failures across entire states and even damage satellites in orbit. To this day, we cannot say conclusively whether or not the Sun's activity also influences the climate on Earth. The above photo was taken on 24 March 2001 in Fairbanks, Alaska, only a few days after a solar eruption.

Photo:
J. Curtis (Univ. Alaska), ACRC

31

Contraction folds on Mercury

The surface of Mercury, the planet that is next to our Sun, is dotted with impact craters. Moreover, long mountain ranges extend across the country in many places. One such ridge, also known as Rupes, stretches from the right-hand edge of this photo to the bottom left. Ridges like these appear on the surface in large numbers. It is thought that they originated very early, when Mercury lost its initial heat and began to shrink. Obviously unable to contract at the same rate, the crust became wrinkled like the peel of an old apple.

Photo:
Messenger Space Probe: NASA
Johns Hopkins University/Carnegie Institution of Washington

32_33

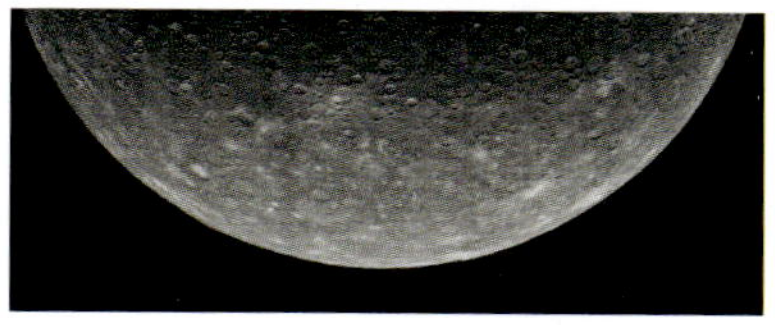

Old and unchanged

Most of the surface of the planet Mercury has remained unchanged for about four billion years, as evidenced by the large number of impact craters. Yet it shows some remarkable phenomena, such as the bright rays that emanate from some craters. They consist of rocky material that was ejected by the impact of a meteorite from the resultant crater. Coming from greater depths, it is fresher than the surrounding surface rock. Mercury is a colorless planet.

Photo:
Mariner 10 space probe, NASA, JPL, USGS

35

Mighty Maat Mons

130_131 One of the most significant insights to emerge from the Magellan mission was that the face of Venus was changed not by meteorite impacts, which are few in number, but by volcanoes. With a height of eight kilometers, Maat Mons is the tallest among them. The ground before it is covered by an extensive lava plain. The picture above was generated from radar data which also contain elevation readings, so that they can be converted into relief maps which may be viewed from any angle on a computer. The colors were derived from photos taken by the Soviet landers Venera 13 and 14. Moreover, the image was vertically stretched by a factor of twenty in order to make the mountain appear more prominent.

Photo:
Magellan space probe, NASA, JPL

36

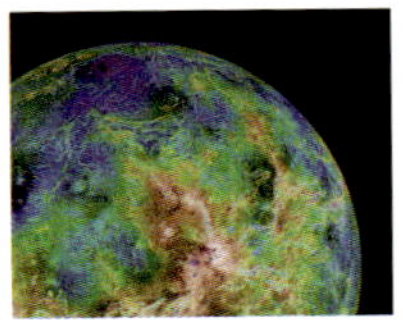

Venus unveiled

At an altitude of about 60 kilometers, Venus is surrounded by a dense blanket of clouds which, consisting of droplets of sulfuric acid, obscures our view of the surface. In the early 1990s, however, the American Magellan probe successfully scanned almost the entire surface of Venus from orbit by radar. From these data, surface relief maps like the one shown above may be generated. In this case, elevations are color-coded so that low-lying plains appear blue while mountain ranges vary between green and white. Rising in the west, the Aphrodite Terra highlands merge with the Atla regio towards the east. The Maat Mons volcano appears as a bright spot somewhat to the east of the center.

Photo:
Magellan space probe, NASA, JPL, USGS

37

Lava plains and volcanoes

Large parts of the surface of Venus consist of hilly landscapes which appear dark on this radar image. Mountain ranges are rare, the largest being Aphrodite Terra which shows up bright in the equatorial region. Maat Mons, the highest volcano on Venus, arises in the east where it appears as a bright spot. Sapas Mons, a somewhat smaller volcano, stands at some distance towards the northwest. Surprisingly, Venus underwent a phase of unusually extensive volcanic activity some 600 million years ago, when viscous lava flooded almost the entire planet. It is not known what caused this global catastrophe. The absence of plate tectonics may well be responsible. On Earth, heat may escape from the interior not only be way of volcanoes but also around the edges of the continental plates. As this was impossible on Venus, the heat dammed up in the interior eventually burst out in an immense volcanic eruption.

Photo:
Magellan space probe, NASA/JPL

38

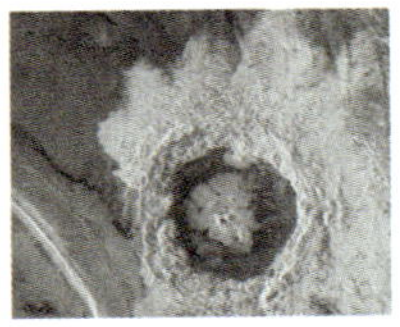

Scars of old meteorite impacts

Not only volcanoes but also meteorites have left their mark on the face of Venus. This radar image shows the Dickinson crater which, measuring 69 kilometers in diameter, is about twice as large as Berlin. Radar data contain information not only about the elevation of a surface but also about its roughness. Smooth regions appear dark, while those strewn with rocks appear bright. Thus, rocks that were thrown up by the Dickinson meteorite impact are bright, while the crater floor of solidified lava appears dark. As most of the ejecta settled to the east of the crater, researchers believe that the meteorite came from the west and hit the surface at an oblique angle.

Photo:
Magellan space probe, NASA/JPL

39

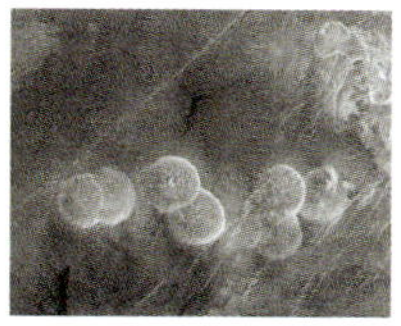

Pancakes

On Venus, researchers have discovered a variety of unique surface formations such as these „pancakes". They are, in fact, circular table mountains which, measuring about 25 kilometers in diameter, rise as high as 750 meters above the surface and display very steep flanks. Their origin is not yet fully understood. Either they are surface bulges thrown up by flows of magma, or highly viscous lava emerging from apertures towered up to form these mountains. The cracks that we see developed when the lava cooled off and fractured.

Photo:
Magellan space probe, NASA/JPL

41

Wide plains and deep canyons on Mars

The surface of Mars is curiously divided: while the northern half is dominated by a wide plain, the south is characterized by high mountains, volcanoes and deep canyons. Distinguishable near the equator, Valles Marineris is the largest rift valley in the solar system. About 4,000 kilometers in length, it is up to 700 kilometers wide and ten kilometers deep in certain places. These dimensions top those of the Grand Canyon in Arizona by a factor of about ten. Valles Marineris originated several billion years ago, when the Martian crust arched up on a regional scale. Two volcanoes rise at the western edge of the planet. What gives Mars a certain resemblance to Earth are the ice caps at the poles, of which the northern one is visible in the picture.

Photo:
Viking Orbiter 1 space probe, NASA, USGS

42

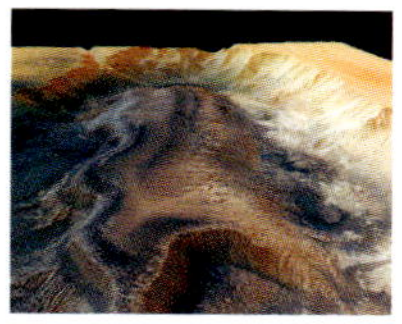

Water on Mars

Searching for signs of former rivers or lakes is a priority goal in Martian research. In the Hebes Chasma valley, such signs have actually been discovered. Hebes Chasma is a depression 300 kilometers in length and up to 130 kilometers wide. In some locations, substances could be identified that must have formed in the presence of water. It may be that these are the sites of former sinter terraces like those you can still see today in the Toscana or at Pamukkale in Turkey. Enclosed all round, Hebes Chasma never had water flowing into it. It is cogitable that part of the basin may once have been covered by ground water. At the center of Hebes Chasma, a giant table mountain rises almost as high as the surrounding country. The rock that forms the walls of the basin and those of the table mountain shows clear evidence of stratification, a window on the history of Mars. Its origin, however, remains enigmatic to this day. Taken by a stereo camera, the picture may be viewed from any perspective.

Photo:
Mars Express space probe, ESA, DLR, FU Berlin (G. Neukum)

43

Deep gorges

Pictured above, Echus Chasma is a ravine around one hundred kilometers long and up to ten kilometers wide that cuts into a high plateau north of Valles Marineris. It may be that a raging torrent once ran through this deep canyon.
The dark material on the surface is probably volcanic ash. In addition, there is a widespread network of bright shallow troughs criss-crossing the plateau, faintly resembling the drainage networks we know from Earth. It is not clear, however, whether the valleys were once filled by precipitation or by water emerging from the ground.

Photo:
Mars Express space probe, ESA, DLR, FU Berlin (G. Neukum)

44_45

Deposits in a Martian canyon

As on Earth, warm periods and ice ages once followed one another on Mars, with each climate change leaving different deposits behind. This is beautifully illustrated by this slope in an enormous gorge named Chasma Boreale which runs through the fringes of the polar cap, nearly cutting it in half. Rearing up as high as two thousand meters, the escarpments of Chasma Boreale show terraced deposits of dust and ice. Probably differing in their ice content, the bright and dark layers indicate climate changes on our neighboring planet. One of the ice fields of the north pole can be seen at the bottom left.

Photo:
Mars Reconnaissance Orbiter space probe, NASA, JPL-Caltech, Univ. of Arizona

46

Dunes in sand and ice

Sand dunes are not only typical of terrestrial deserts, they also dominate large segments of the Martian surface. Located in the vicinity of the Martian north pole, some of the dunes pictured above display the V-shape that is typical of the sickle-shaped dunes on Earth that are known as barchans. On Earth, they are shaped by the wind, but atmospheric pressure on Mars amounts to no more than one percent of that on Earth. Under these conditions, it must have taken many thousands or even millions of years for these barchans to take shape. Besides, ice deposits can be seen around their fringes – remnants of the previous Martian winter.

Photo:
Mars Reconnaissance Orbiter space probe, NASA, JPL-Caltech, Univ. of Arizona

47

At the rim of the Victoria crater

Large parts of the Martian surface are marked by impact craters. Measuring 730 meters wide by 70 meters deep, the Victoria crater shown above is located close to the equator in a plain called Meridiani Planum. It is a particularly interesting object for scientists because several strata can be distinguished on its slopes. These are probably layers of sediment deposited during various climatic phases. Dunes of fine dust have formed at the bottom of the crater.

In September 2006, the American Mars rover Opportunity reached the Victoria crater, investigated it for several months and went partway down the slope. Its investigations permit drawing conclusions regarding the development history of Mars. This extremely detailed photo even shows Opportunity at the edge of the crater (bottom left), at about 18.30 h.

Photo:
Mars Reconnaissance Orbiter space probe, NASA/JPL-Caltech, Univ. of Arizona, Cornell, Ohio State University

49

Jupiter and Io

This composite photograph shows what Jupiter and its nearest moon, Io, would look like from a distance of about one million kilometers. Jupiter's image is composed of infrared photos taken at various wavelengths, a technique which permits looking more or less deeply into its atmosphere. In this false-color image, the Great Red Spot appears whitish, clouds and haze in the upper cloud layer are bluish, and deeper clouds appear reddish.

Conversely, Io appears in approximately true colors. In the region of its north pole, the cloud of an eruption can be seen rising more than 300 kilometers, shining bluish in the light of the Sun. Tvashtar, a volcanic hotspot, can be distinguished in the shadow below the cloud as a reddish glowing dot.

Photo:
New Horizon space probe, NASA, Johns Hopkins University, Southwest Research Institute, Goddard Space Flight Center

50

The tempestuous giant

This is the most detailed color image ever produced of the entire planet Jupiter. It was generated by superimposing images made with red, green and blue color filters on a computer. The size of the smallest details shown here is 60 kilometers. Despite its enormous dimensions, the planet takes only ten hours to turn on its axis, keeping the outer layers of the atmosphere in the vicinity of the equator whirling around at a speed of more than 40,000 km/h. This is the reason why Jupiter's clouds are distended into colored bands that can be identified in mid-range amateur telescopes. Their colors reflect different substances in the atmosphere, mainly ammonia, ammonium hydrogen sulfide (NH_4SH) and water.

Photo:
Cassini space probe, NASA, JPL,
Space Science Institute

51

The Great Red Spot

The most conspicuous and famous phenomenon in Jupiter's atmosphere is the Great Red Spot in the southern hemisphere, a gigantic cyclone measuring about 22,400 by 12,000 kilometers. Two Earths would fit in it comfortably. This mighty storm has been in existence for 300 years at least.

To the north and south, winds moving in opposite directions circle the Great Red Spot. Moving between these flows, the vortex rotates around its axis in six days. Occasionally, it swallows up one of the white cyclones in the vicinity which, although smaller, still have the dimensions of a continent on Earth, probably using their momentum to wind itself up like a top. The cause of its reddish-brown coloration is not clear. It may be due to phosphorus that is released by molecules as they are broken up by lightnings or the sunlight. Another likely alternative is that it is caused by organic molecules wafted up from the deeper layers of the atmosphere. The colors in this image are approximately those which a human eye would see.

Photo:
Voyager 1 space probe, NASA, JPL

52_53

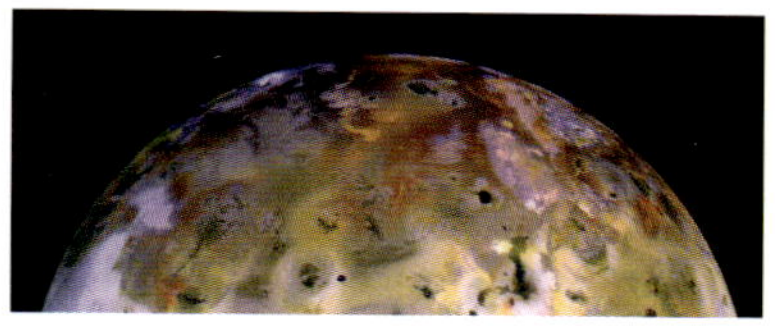

Io – the body with the greatest volcanic activity in the solar system

The greatest volcanic activity in our solar system is to be found not on Earth but on Jupiter's innermost moon, Io. Between 1995 and 2003, the Galileo space probe observed around 120 volcanoes spewing gas and dust up to 400 kilometers high. Within the space of 100 years, they eject enough matter to cover the entire globe with a layer about ten centimeters thick, so that the surface changes constantly. Up to 300 kilometers long and as hot as 1,500 degrees centigrade, flows of lava consisting of sulphurous material and molten silicates roll across the surface. The moon's reddish-yellow color is also caused by various sulfate compounds. Now, where does the energy come from that is needed for such volcanic activity?

It is generated by the tides which Jupiter causes in Io. These deform the moon, periodically kneading its outer shell, the mantle, like dough. The heat thus generated is enough to produce liquid magma. The tidal force exerted by Jupiter on Io is three hundred times greater than Earth's influence on the Moon.

Photo:
Galileo space probe, NASA, JPL,
University of Arizona

54_55

Ice fissures on Europa

Close-ups of the surface of the Jovian moon Europa show a pattern of reddish-brown trenches criss-crossing its blue ice shell. Two to three kilometers wide and several hundred kilometers long, these fissures probably formed when the ice cover cracked under the influence of Jupiter's powerful tidal forces. Slushy ice emerging from the cracks settled along their edges, forming a brownish deposit. At the same time, a mixture of gas, rock and ice shot out from the cracks. As these eruptions went on, the increasingly pure and white ice they produced formed embankments to the right and left of the cracks that reach a hundred to two hundred meters in height. We do not know whether such activities are still going on on Jupiter today.

Photo:
Galileo space probe, NASA, JPL, DLR

56

Ganymede

Ganymede is the largest moon in the solar system. Its diameter of 5,268 kilometers is greater even than that of the planet Mercury. Its density is only about twice that of water, the obvious conclusion being that it mostly consists not of rock but of water ice. The thick icy crust that forms its surface features numerous cracks and fissures, indicating that the moon must have been geologically active at least during its early history, and that its surface has been extensively distended and compressed. The large, nearly circular dark areas were caused by the impact of enormous meteorites. Surprisingly, Ganymede has its own magnetic field that is probably generated by an ocean of warm saline water moving about in its interior.

Photo:
Galileo space probe, NASA, JPL

57

Callisto – an ancient icy world

Of the four moons discovered by Galileo in 1609, Callisto is farthest away from Jupiter. Its surface is an icy desert, frozen long ago and partially covered with rock dust. Its salient feature is an immense number of impact craters, from which scientists conclude that any geological activity probably ceased more than four billion years ago. Named Valhalla, the giant impact crater shown here measures more than 3,000 kilometers in diameter, which probably makes it the largest of all. On icy and rocky moons, impact craters differ in their appearance. The structural details of the crater appear in the ice like a carbon copy. Valhalla originated at least three million years ago when a giant meteorite hit Callisto, tearing up numerous concentric rift structures. The white ice catapulted from the ground by later impacts contrasts with the ancient dark surface. It may be that an ocean of viscid slushy ice exists below a layer of ice and rock several hundred kilometers in thickness.

Photo:
Galileo space probe, NASA, JPL, DLR

59

Surreal Saturn

In this photo, Saturn appears to be floating in space like a translucent sphere. As the Sun was standing behind Saturn when this image was taken, its light shines through the rings from below. It also shines through the planet's outer atmosphere, making it appear as a bright ring. Enough light is deflected to light up the rings on the night side of Saturn, so that we can see them although they are in the planet's shadow. In this slightly overexposed picture, moreover, researchers discovered two hitherto unknown rings. What is most fascinating about this picture, however, is the small white spot that appears outside the main ring system at about 10:00 h. It is our planet Earth. At the time when this picture was taken, it was more than one billion kilometers away from Saturn.

Photo:
Cassini space probe, NASA, JPL,
Space Science Institute

60_61

Born from a collision

Saturn's large main rings can be seen even in a low-powered telescope. However, researchers have discovered a number of additional, very fine rings with the aid of space probes. The main ring system measures 280,000 kilometers or twenty times the Earth's diameter, yet its thickness is only 30 meters. If we were to reduce the ring disk to the size of two football fields, its thickness would shrink to one hundredth of a millimeter – thinner than a sheet of paper. This makes the rings of Saturn the flattest known object in the universe. How did they originate?

According to one theory, there was once a moon in the space that is now occupied by the rings. This moon was probably hit and broken up by an asteroid. The remaining rock fragments were then ground into dust by successive asteroid impacts and mutual collisions. Spreading out over several orbits, the resultant particles formed today's ring system. As ring particles may also agglomerate into small moons, a dynamic cycle of growth and decay is operating in the rings.

Photo:
Cassini space probe, NASA, JPL,
Space Science Institute

62

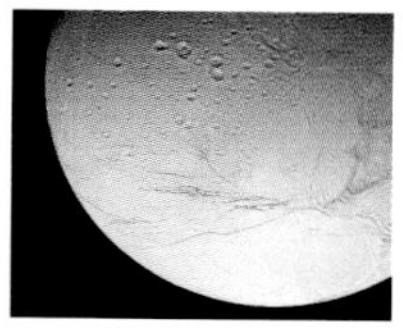

Fountains on Enceladus

Saturn's moon Enceladus is one of the most interesting bodies in the solar system. In its south polar region, jets of water ice rise several hundred kilometers above the surface. Some particles even leave the moon's gravitational field and spread out along its orbit, thus forming what is called Saturn's E-ring. Moreover, the fountains that rise from fissures have been shown to contain organic molecules. It may be that an ocean consisting of either water or a mixture of water, ice and rock extends below the thick ice crust.

It is Saturn that provides the energy for the „cryovolcanism" mentioned above. Its powerful tidal forces flex and heat the interior of Enceladus, producing reservoirs of water ice close below the surface. As these expand, the pressure inside them rises until the water bursts out through the crust.

Photo:
Cassini space probe, NASA, JPL,
Space Science Institute

63

Titan above the ring plane

Seen here floating above the plane of Saturn's rings, Titan is one of the most exotic heavenly bodies in our solar system, and with a diameter of 5,150 kilometers it is also its second largest moon. What is unique about it is its dense atmosphere, a feature not found on any other moon. It is a veritable laboratory for organic chemistry. The outer layer of the atmosphere consists of thick smog which obstructs our view of the surface.

Consisting almost exclusively of nitrogen and methane, the atmosphere reaches a pressure of about 1.5 bar on the surface, where temperatures around minus 180 degrees centigrade prevail. Under these conditions, methane plays a role similar to that of water on Earth: in the atmosphere, methane clouds float from which a fine drizzle falls that collects in lakes and rivers. In the high northern latitudes, there are giant seas filled with liquid methane and ethane – pit gas. With the exception of Earth, this makes Titan the only body in the solar system with a liquid substance existing on its surface.

Photo:
Cassini space probe, NASA, JPL, Space Science Institute, S. v. Vuuren, outsideinthemovie.com

65

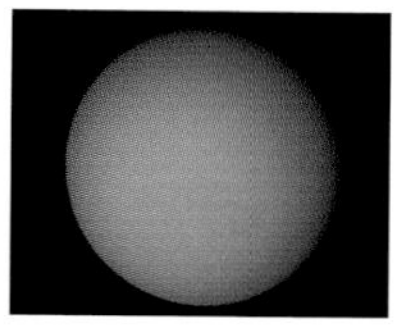

Uranus – the blue sphere

The planet Uranus appears to us as a uniformly blue sphere. Its extensive atmosphere consists largely of hydrogen and helium. At the low local temperature of minus 215 degrees centigrade, methane freezes out, forming haze particles which give the planet its blue color.

At greater depths, the atmosphere grows denser until it behaves like a liquid and even assumes metallic properties. Lower down, the proportion of silicates and iron increases until we reach a core of rock and iron weighing about four times as much as Earth.

Surprisingly enough, the planet's spin axis is inclined 98° against its orbital plane. Rolling along like a cartwheel, the planet rotates backwards against the direction of its orbital movement. No conclusive explanation for this phenomenon has been found so far. It may be that the planet acquired a „list" when it collided with a large heavenly body. This inclination is the reason why, during the planet's 84-year orbit, each of its poles faces towards the Sun about half the time and away from it for the other half, so that a day there lasts around 40 years.

Uranus and Neptune are the only planets that cannot be observed in the sky with the naked eye. This is why they were only discovered in modern times, when telescopes were available. It was William Herschel, the astronomer, who almost accidentally discovered Uranus on 13 March 1781 while observing stars in his garden.

Photo:
Voyager 2 space probe, NASA, JPL

67

A last look back

After the Voyager 2 space probe had passed the outermost planet, Neptune, in August 1989, it turned its camera back for this image of the south polar region, sparsely illuminated by the Sun. This marked the end of one of the most successful interplanetary space missions of all time, twelve years after its launch. On its „Grand Tour", Voyager 2 had passed Jupiter, Saturn, Uranus and Neptune. The two outermost planets have not been visited by a space probe since.

Photo:
Voyager 2 space probe, NASA, JPL

68_69

Triton and Neptune

136_137 This composite picture shows Neptune behind its moon Triton. On Neptune, violent storms are raging, and white clouds of methane ice race hither and thither at up to 600 km/h. In 1989, researchers discovered a dark spot the size of Earth that recalls the Great Red Spot on Jupiter. This funnel-shaped cyclone is surrounded by white cirrus clouds of methane ice which are also known as scooters.

Although Neptune's moon with its diameter of 2,760 kilometers is somewhat smaller than our Moon, the diversity of its surface structures is amazing. A large area surrounding the south pole is colored pink, probably because it is covered by methane and nitrogen ice. The rocky ground to the north, on the other hand, appears greenish-grey. In addition, there are dark fountains to be seen that appear to shoot up from the ground, leaving behind oval-shaped black traces on the surface that measure up to 150 kilometers in length. It is supposed that these are geyser-like eruptions which occur when subsurface bubbles filled with nitrogen gas expand in the sunlight until they finally explode.

Photo:
Voyager 2 space probe, NASA, JPL, USGS

71

Comet Hale-Bopp

In April 1997, the Hale-Bopp comet appeared in the sky and developed into one of the brightest objects seen in the 20th century. Although the comet's body measured no more than 40 kilometers in diameter, its tail was 200 million kilometers long at the peak, a distance greater than that between the Earth and the Sun. Moreover, Hale-Bopp displayed very clearly a phenomenon called a twin tail. The white tail consists of minute dust particles that are propelled into space by the sunlight. The bluish tail consists of electrically charged atoms (ions) and molecules that are blown away from the comet by the solar wind.

Photo:
Corbis, A. Horowitz

72

Ida and Dactyl

Ida (shown above) and Gaspra were the first asteroids targeted by a space probe. With a length of 60 kilometers, Ida is one of the larger asteroids. The discovery of a small, 1.5-kilometer moon orbiting Ida at a distance of no more than 85 kilometers came as a complete surprise. It was probably torn off when Ida collided with another asteroid and has been orbiting its parent body ever since, unable to leave its gravitational field. If astronauts were to land on this miniature moon called Dactyl, they would find themselves in a weird world. They would be able to circumambulate the entire body within a single hour, and they would feel as light as a feather, almost weightless. The gravity on its surface is so low that throwing a pebble into space would be no problem. Above, Ida would float in a black sky, appearing as large as a terrestrial football stadium seen from a hundred meters away.

Photo:
Galileo space probe, NASA, JPL, USGS

73

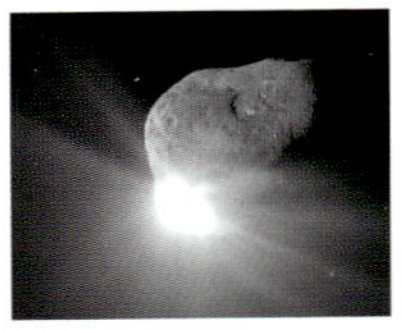

Shooting at a comet

In the early morning of July 4, 2005, a copper projectile struck the comet Tempel 1 at a speed of 37,000 km/h. Weighing 365 kilograms, it had been ejected the previous day by NASA's Deep Impact space probe. This sensational experiment was launched by researchers in order to learn more about the inner structure of comets. During the impact, the Deep Impact carrier flew past the comet, which measures 7 kilometers in length and 4 kilometers in thickness. 55 terrestrial and several space telescopes were pointing at the comet at the time. Evaporating completely on impact, the copper projectile produced a crater which probably measures 100 meters in size. When they analyzed the material that was ejected from the comet, researchers discovered a number of familiar substances like water and carbon dioxide as well as more complex organic compounds, such as methanol and formaldehyde.

Photo:
Deep Impact space probe,
NASA/JPL-Caltech/UMD

74

Comet McNaught

In January 2007, the comet McNaught developed into one of the brightest objects of the last few decades. In the photograph, it barely clears the horizon above the bright lights of Santiago de Chile. The tail of this comet took on an unusual fanlike shape caused by jets of gas and dust which, erupting from the core, shone in the sunlight.

Photo:
S. Guisard, ESO

75

The Barringer crater – evidence of a cosmic disaster

Asteroid impacts have crucially influenced the development of life. The 1.5-kilometer Barringer crater in Arizona is the „scar" left behind by such an event. About 50,000 years ago, an iron asteroid around 50 meters in size struck here, developing an explosive power more than a thousand times greater than that of the atomic bombs dropped on Hiroshima and Nagasaki. Scientist calculate that everything within a radius of 20 kilometers was destroyed, and all life perished within a distance of 4 kilometers.

65 million years ago, Earth was struck by a rocky asteroid of 8 to 10 kilometers with such force that the sky was darkened for years by clouds of dust and the smoke of burning forests. Scientists suppose that the extinction of the dinosaurs was one of the consequences of this event. The crater caused by the impact is thought to be located on Mexico's Yucatan peninsula. Asteroid impacts and their global consequences probably influenced the evolution of life in earlier times as well.

Photo:
S. Seip – astromeeting.de

77

„Look at that picture!"

Frank Borman, James Lovell and William Anders were the first human beings to fly to the Moon on Apollo 8 in December 1968. On Christmas Eve, they saw the other side of the satellite for the first time, and they were profoundly moved when the Earth rose above the horizon of the Moon. „Oh my God", Anders suddenly exclaimed, „Look at that picture over there! Here is the Earth coming up." Although the astronauts did not really have the time to take photographs, Anders was so fascinated that he reached for the camera and shot two color pictures of the rising Earth. They are now among the most famous photographs of the Apollo era. There is probably no other picture which shows the contrast between Earth, the island of life, and the dead landscape of the Moon as impressively as this one.

Photo:
Apollo 11, NASA

78

The back of the Moon

Taken by the Apollo 11 crew, this photo of the back of the Moon shows a surface riven by impact craters. Generally speaking, the „seas" on the back of the Moon are not as large as those on the front. Why this should be so, we do not know. While Michael Collins was orbiting the satellite in Columbia, the parent craft of Apollo 11, Neil Armstrong and Buzz Aldrin went on to become the first human beings to land on the Moon. On 21 July 1969, at 3 hours, 56 minutes and 20 seconds central European time, Armstrong's boots stirred up the dust which had lain undisturbed on the Moon's surface for billions of years. The Moon rocks that were brought back to Earth by the astronauts of the seven Apollo missions have been analyzed in numerous laboratories around the world. They crucially influenced our present-day theory about the origin of the Moon.

Photo:
Apollo 11, NASA

138_139

79

Above the ocean of air

This photo of the sun setting above the Pacific was shot from an altitude of 378 kilometers by an astronaut on board the International Space Station (ISS). It demonstrates the different manifestations of our elixir of life, water. More importantly, it gives us an inkling of how tender our atmosphere really is. If we imagine the Earth reduced to the size of an apple, the atmosphere would be thinner than its skin. Three quarters of all the air we have is concentrated within the first 10 kilometers above the surface. This is the site of all essential meteorological phenomena. All the exhaust gases produced by vehicles, industrial plants, power stations and households are emitted into a layer of air whose thickness approximately corresponds to the distance between the city centers of Oberhausen and Duisburg.

Photo:
ISS, Expedition 7 Crew, EOL, NASA

80_81

The Moon – a child of Earth

According to the theory that is most favored today, the Moon formed early in the history of the solar system when a body the size of Mars collided with the still-young Earth. Part of the material that was blown away entered into an orbit around Earth, where it agglomerated to form the Moon. The above picture combines image data from two weather satellites with an elevation relief map. The latter was vertically stretched by a factor of 50 to make mountain ranges appear more plastic. The contrast between green areas of vegetation and yellow arid zones is especially clear. The hurricane Linda can be seen raging off the western coast of Mexico, and the atmosphere rests on the Earth's globe like a fine layer of hoarfrost.

Photo:
R. Stockli, A. Nelson, F. Hasler, Laboratory for Atmospheres, Goddard Space Flight Center, Data from NASA, NOAA, USGS

82

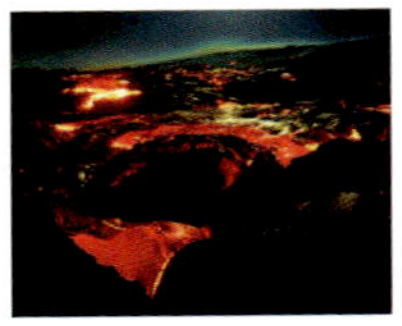

Lava flows on Hawaii

Volcanoes constitute one of Earth's most impressive natural phenomena, having left their mark on the surface over billions of years to the present day. Volcanism also provides tangible proof that the interior of our planet is hot. On Hawaii, liquid lava from the Earth's interior rolls down to the sea in wide streams.

Hawaiian volcanism is a special phenomenon, for we are looking at a chain of volcanoes that formed one after the other when the Pacific continental plate moved across a subterranean magma chamber (hotspot) at a rate of 50 kilometers per one billion years. Like a cutting torch, successive blasts of magma moved up through the plate, producing volcano after volcano. Plate tectonics is the main cause of volcanism all over the Earth.

Photo:
G. Brad, Lewis, SPL Agency Focus

83

A chain of volcanoes

Like the bones of a spinal column, Hawaii's volcanoes rear up from the Pacific. The main island of Hawaii in the extreme southeast and the small islands in the extreme northwest are separated by around 3,500 kilometers. Further along, the chain turns towards the north, as we can see at the western edge of the picture. The Mauna Kea volcano rises 4,205 meters from the sea, while its base rests 5,400 meters under water. This makes it the tallest mountain on Earth. Hawaii's volcanoes developed successively over a period of about 70 billion years. Today, active volcanoes exist only at the southeastern end of the chain. Loihi, the most active one, is submerged in the sea. The elevation model shown above is computer-generated.

Photo:
NGDC

84

The Alps – a crumple zone between continental plates

Europe's highest mountains, the Alps, are an impressive product of plate tectonics: two colliding continental plates have been thrusting up this range. When the process began about 30 million years ago, the rate at which the mountains rose was as high as 5 millimeters per year. Today, the process is weakening, so that the Alps only grow by less than half a millimeter per year. On the left-hand side of this satellite image you can see the Montblanc massif whose rugged conical peaks were shaped by the last ice age.

Photo:
Landsat 7 satellite, University of Maryland, GLFC, DLR

85

A deep rift

Resembling an outsize scar, the San Andreas Fault extends over 1,300 kilometers through Mexico and California. It marks the seamline between two continental plates that move past each other at an average rate of 6 centimeters per year. Again and again, earthquakes relieve the tension that builds up in the process, of which the quake that destroyed San Francisco on 18 April 1906 was particularly devastating.

In our solar system, plate tectonics exists nowhere but on Earth. Fitting together like a global puzzle, continental plates float on an elastic layer called the astenosphere. Wherever two plates drift apart, long rifts open up in which hot crust material rises, pushing continents apart by as much as ten centimeters per year. In other places, one plate slides under another and dives into deeper layers. There, mountain ranges like the Himalayas rise up. By circulating rocky material, plate tectonics integrates it into a never-ending cycle.

Photo:
NHPA, K. Schaffer

86

Black smokers – life under extreme conditions

Until a few years ago, biologists assumed for a fact that sunlight was indispensable for the existence of life. Therefore, the discovery of black smokers was a spectacular event. These hot springs are found in the depths of the ocean where no sunbeam ever penetrates. Especially along the mid-ocean ridge, where continental plates are being pushed apart by rising magma, a mixture of minerals, organic substances and metallic compounds bubbles up from the interior of the Earth. The flues were named after the black flakes which the metallic compounds form as they fall out in the icy seawater. Forming the first link in the exotic food chain of the symbiotic communities living there, bacteria extract energy mainly from the chemical conversion of hydrogen sulphide. It was there that the current holder of the record in „high-temperature living" was discovered. Bacteria of the strain 121 are best suited by ambient temperatures of 121 degrees centigrade.

This life in darkness, nourished by the inner heat of our planet, has led some scientists to speculate whether there might not be life in the oceans that are supposed to exist in the interior of the Jovian moon Europa.

Photo:
Stephan Low Company, Rutgers Univ., Giant Screen Motion Pictures

87

„Moon landing'"

The Swedish photographer Lennart Nilsson has named this image 'Moon landing'. Yet his camera was aimed not into space but into the interior of the human body, recording the moment when a fertilized egg cell aged between 5 and 6 days reached the womb. Using a camera technology developed by himself, Nilsson explored a universe of baffling beauty: the depths of our body. In it, the miracle of human development has been enacted for millions of years with amazing immutability, forming part of the eternal cycle of life and death. There are some photos that crucially inform our view of nature. They include photos of our blue planet taken from an Apollo spacecraft as well as Lennart Nilsson's photographs of life in the making.

Photo:
L. Nilsson

88

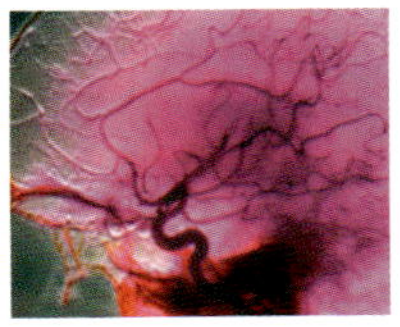

Brain arteries

140_141

The human brain represents the ultimate stage in the evolution of life on Earth. It is estimated to contain about one hundred billion nerve cells linked by about one hundred trillion synapses, forming the most powerful computer we know. Although it only accounts for 2 percent of our body mass, about 20 percent of our entire blood volume flows through the brain every minute, supplying it with energy. This purpose is served by a ramified network of blood vessels whose dendritic structure stunningly resembles that of a river delta.

It appears miraculous that bodies as complex as that of Earth and its inhabitant, man, should have developed from the chaos of extremely hot gas produced by the Big Bang. For this cosmic evolution, the synthesis of elements in stars was indispensable as it provided the universe with vital substances like carbon and oxygen. Man is indeed made of stardust.

Photo:
Alain Pol, SPL, Focus Agency

89

River arteries

The delta of the Kaladan river in Myanmar (Birma) graphically shows how dendritic structures evolve by self-organization. On its way from source to mouth, the river carries enormous loads of sand and mud with it. When its current slows down and its transport capability declines in consequence, sediments pile up, obstructing the river and forcing it to seek another bed for itself. Repeated any number of times, this process ultimately produces the numerous branches of a delta whose shape, size and number are in constant flux. The soil between the branches of a delta is often very fertile. In this radar image, mangrove forests appear dark green and watercourses black.

Photo:
Space shuttle Endeavour,
Radar SIR-C/X-SAR, DLR

90

The spaceman

This is the title given by Swedish photographer Lennart Nilsson to this picture of an embryo in the womb. Eleven weeks old and six centimeters in size, it appears to exist in weightlessness like, for example, an astronaut on board the International Space Station.

Photo:
L. Nilsson

91

Lost in space

Floating high above the Earth in the blackness of space, astronaut Bruce McCandless II appears lost and lonely. When this photograph was taken in February 1984, McCandless and four other astronauts had just flown into orbit in the American space shuttle Challenger to drop off two satellites. He is seen here testing the Manned Maneuvering Unit, a kind of rocket chair that can be controlled by nitrogen-powered jets. Driving that chair, McCandless moved one hundred meters away from the space shuttle and returned safely.

Photo:
NASA

92_93

The last man on the Moon

In December 1972, the astronauts of Apollo 17 enjoyed the comfort of an electric vehicle for traveling around the Moon. Harrison Schmitt, a geologist and the first and hitherto only scientist on the Moon, appears in this picture against the backdrop of the barren Moon landscape. Schmitt is walking through a field littered with rocky debris near Camelot Crater in the Taurus-Littrow valley at the eastern edge of Mare Serenitatis, a landscape bordered by mountains two and a half thousand meters high. Schmitt and his teammate Gene Cernan were the last men to visit the Moon, which they left on 14 December 1972.

Photo:
Apollo 17, NASA, JSC

95

Explosion in the Unicorn

Early in 2002, a star designated as V838 Monoceros lit up unexpectedly in the Unicorn constellation. 20,000 light years distant, this extremely massive star now nearing the end of its life was swollen into a red supergiant.
When it finally erupted, its luminosity increased to 600,000 times the power of our Sun, making it the most brilliant star in the Milky Way for a brief period. If this supergiant had been in the place of our sun, its outer fringes would have almost touched Saturn's orbit. V838 Monoceros is surrounded by several shells of gas and dust ejected before its eruption. Among other things, the shells contain heavy elements like carbon and oxygen which are now available to the next generation of stars and planets.

Photo:
Hubble Space Telescope, NASA, ESA,
Hubble Heritage Team STScl, AURA

96_97

Growth and decay

Named Pismis 24, this star nursery in the Scorpio constellation is about 8,000 light years away from Earth. Some young stars illuminate from inside the cloud of dust that screens them. Towards the right, this formerly closed cocoon has opened under the influence of the energy-rich radiation and particle winds emitted by these stars. The brightest star that can be seen outside the cloud on the right is one of the most massive in our galaxy. It is about one hundred times as heavy as our sun. Stars of that size are extremely wasteful with their fuel, exploding in supernovas only a few million years after their birth. This giant, too, is bound to die in the not-too-distant future, blowing most of its substance into space and, by the same token, closing the cycle of growth and decay, for its material will be consumed in the making of other stars in the future.

Photo:
Hubble Space Telescope, NASA, ESA,
J.M. Apellÿniz (Instituto de astrofisica de Andalucia, Spain), D. de Martin

98

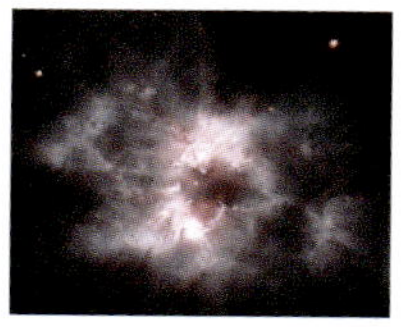

The Insect nebula

About 4,000 light years distant, this planetary nebula in the Puppis constellation is occasionally called Insect nebula because of its unusual shape. A white dwarf sits at its center, a star which, having burned up its fuel and contracted to the size of Earth at the end of its life, is now slowly cooling off. Before it turned into a white dwarf it threw off its outer atmosphere in several phases, forming what we see today as the Insect nebula. Our sun too will turn into a white dwarf in about six billion years. However, we cannot say whether, before that happens, it will also throw out part of its atmosphere, which would then appear from afar as a planetary nebula.

Photo:
Hubble Space Telescope, H. Bond (STScl),
R. Ciardullo (PSU), NASA, ESA

99

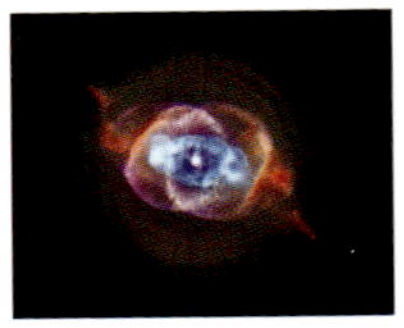

Gas shells like tree rings

The planetary nebula called the Cat's Eye is one of the most complex of its kind. Nothing to do with planets, planetary nebulas are clouds of gas and dust thrown off by a dying star. In the outer reaches of the nebula, several concentric rings can be distinguished that were caused by eruptions of the central star which, probably occurring at 1,500-year intervals, created the gaseous shells that are still moving outward today. Astronomers observe these shells to reconstruct the last 10,000 years of the star's history, like biologists studying the annual rings in trees. With the aid of spectrography, they were able to demonstrate the presence of vital elements like carbon, oxygen and nitrogen in the gas.

This image was generated by superimposing two pictures taken in visible light and the X-ray band. The X-rays are emitted by extremely hot gas that appears blue-violet.

Photo:
Hubble and Chandra Space Telescopes, NASA, CXC, SAO, STScl

100

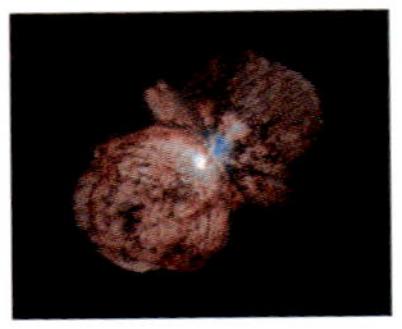

The next supernova in the Milky Way

In 1837, British astronomer John Herschel observed that Eta Carinae, a star in the Carina (keel) constellation, was slowly growing brighter. Six years afterwards, it was the second most brilliant star in the sky after Sirius. Later, its light dimmed, and today it is no longer visible with the naked eye. At the time, Herschel asked himself, „What conclusions should we draw regarding the comfort or habitability of a system whose supply of light and heat depends on so unsafe a source?" Today, it has become clear that there cannot be any planet orbiting Eta Carinae on which conditions favor life. With a mass about one hundred times that of the sun, it is one of the most massive and unstable stars in the Milky Way.

When its luminosity erupted in the last century, the star threw off a carbon-rich gaseous shell that is now spreading outward at 700 kilometers per second. 8,000 light years distant, this giant star has reached the final stage of its life and will explode within the next 100,000 years.

Photo:
Hubble Space Telescope, Jon Morse (Univ. of Colorado), NASA

101

A chemical factory dating back to 1045

In 1045, a star in the Cancer constellation began to shine so brightly that it could be seen even in the daytime, an extraordinary event which was duly recorded in drawings by Asian astronomers. In the same spot, modern astronomers have discovered a cloud of gas that bears witness to the explosion of a star, a supernova.
It was proven to contain a wealth of chemical elements like carbon, oxygen, nitrogen, iron, neon and sulphur. Bred in the interior of the star, they were ejected into space when it blew up. When the supernova exploded, only the outer part of the star was blown off, whereas its center collapsed into a neutron star, the most compact heavenly bodies found in nature. They contain the concentrated mass of our Sun in a ball measuring no more than 20 kilometers in diameter. A teaspoon full of its matter would weigh several hundred million tons on earth.

Moreover, this neutron star at the center of the Crab nebula rotates 30 times per second. The fastest known neutron star even rotates 716 times every second! The latter object is about to be torn apart by centrifugal forces.

Photo:
Very Large Telescope, ESO, FORS Team

102_103

Sulphur in the Witch's Broom

Whenever a star explodes, the resultant cloud spreads at breakneck speed. Its gas grows thinner and thinner until it blends with the ubiquitous interstellar gas and becomes indistinguishable from it. Already more than 15,000 years old, the Cirrus nebula occupies a space in the sky that is large enough to contain six full moons. The part of it that is shown above is called the Witch's Broom nebula.

The picture was generated by superimposing three photos taken through special filters to distinguish various elements: hydrogen appears red, oxygen blue and sulphur green. The bright star occupying the foreground is nothing to do with the nebula.

Photo:
Wisconsin Indiana Yale Observatory & NOAO, T.A. Rector (Alaska U.), AURA, NSF

104_105

An eagle in its nest of stars

The dust cloud that occupies the center of this picture looks like an eagle. This Eagle nebula was made famous by a photo taken by the Hubble space telescope in which it appeared as „the pillars of creation". Showing the surroundings of the nebula, the photograph demonstrates that the eagle floats before a giant „cave" opening up within a cloud of dust. The numerous new stars that have formed in it appear violet. The cave was created by the radiation and the powerful particle winds emitted by these stars. Thanks to a special photographic method, various elements can be distinguished: hydrogen appears green, oxygen blue and sulphur red.

Photo:
Spitzer Space Telescope, T.A. Rector, B.A. Wolpa, NOAO, AURA, NSF

106_107

The Sombrero galaxy

Until about 80 years ago, many astronomers were firmly convinced that our galaxy is the only galaxy in the universe. Today, we know that these star islands number more than a hundred billion. 30 million light years away, the Sombrero galaxy is one of them. It is in fact a spiral galaxy, but we cannot see its arms because we are looking at it almost precisely edge-on. It displays two remarkable characteristics: first, its center is surrounded by a very large sphere which incorporates billions of stars. Astronomers suppose that the center itself is occupied by an enormous black hole of a mass corresponding to about one billion suns. Second, the entire galaxy is surrounded by a band of dust clouds. Observations in the infrared have shown that an immense number of stars are forming in it. This „dust garland" approximately corresponds to a similar band in the Milky Way that is permeated by dark clouds.

Photo:
Hubble Space Telescope, AURA, Hubble Heritage Team STScl, NASA

108

Starbirths boom as galaxies collide

At the high velocities with which galaxies move through space, collisions are inevitable. In such an accident, nothing happens to the stars themselves because they are extremely sparsely distributed. But when two galaxies approach, their gravity fields interfere with one another. Powerful tidal forces arise that draw masses of gas and dust out of the two star systems, forming them into long arms. At the same time, giant dust clouds whirl around chaotically within each galaxy, colliding and condensing. This is why the star birthrate virtually explodes when galaxies collide. In a frontal collision, the two galaxies will merge into a new star system.
300 million light years away from us, the two galaxies shown here have been nicknamed „the mice" because of their tails.

Photo:
Hubble Space Telescope, NASA, ESA,
H. Ford (JHU), G. Illingworth (USCS/LO),
M. Clampin (STScl), G. Hartig (STScl),
ACS Science Team

109

Cosmic islands

If we were to leave our galaxy in a spaceship to view it from a great distance, it would look about the same as this spiral galaxy called NGC 7331, which is about 50 million light years distant. You can see the bright center, whose middle is very probably occupied by a supermassive black hole, as well as the curved spiral arms that are permeated by dense clouds of dust – star nurseries. All in all, this galaxy holds more than a hundred billion stars. The cluster of apparently smaller galaxies above NGC 7331 is around ten times farther away from us. At even greater distances, yet more galaxies can be seen which this photograph shows in all their loneliness and majesty.

Photo:
Sierra Remote Observatories, D. Goldman

PUBLICATION DATA

EXHIBIT LOANS

Argelander Institute of Astronomy, Bonn

Carl Zeiss AG, Oberkochen

CHRONUS-Manufaktur/Hiller & Müller GbR

Exhibition of German Astronautics Morgenröthe-Rautenkranz

German Aerospace Center (DLR)

EADS Astrium

European Space Agency (ESA)

Kassel, Hessen-Kassel Museum Landscape

Schott AG

City of Bochum – Bochum City History Center – City Archive

German Museum of Technology, Berlin

Münster University, Institute of Planetology

Wuppertal Clock Museum

Zeiss Planetarium, Bochum

Prof. Dr. Gerhard Neukum, Free University of Berlin

„Out of this World – Wonders of the Solar System" will be shown at the Oberhausen Gasometer from 1 April 2009 to 30 December 2010. The exhibition is organised jointly by the German Aerospace Center (DLR) and Gasometer Oberhausen GmbH to mark the astronomical year 2009.

EXHIBITION

Director
Jeanette Schmitz

Idea
Jeanette Schmitz, Sabine Göge

Concept
Prof. Dr. Peter Pachnicke, Wolfgang Volz
Wissenschaftliche Beratung: Ulrich Köhler

Moon sculpture
Wolfgang Volz
Music: Ulrich Sende

Space images and Moon maps
DLR Institute of Planetary Research
Ulrich Köhler, Susanne Pieth, Marita Wählisch

Poster and flyer design
Uwe Loesch

Press and public relations
Thomas Machoczek, DLR Communications

Scientific assistants
Dr. Thomas Bührke, Daniel Fischer, Judith Schüller, Klaus Gering, Martin Meyer, Bernd Pachnicke, Sascha Bast, Marco Dorr, Kirsten Galert

Photo production
special:effexx

Light design
start.design

Exhibition furnishings
Kalle Krause, Tischlerei Deflieze, Jost Elektrotechnik

Technical support
Norbert Greser

Video editor
Thomas Wolf

Layout
Gitta Hülsmann

Climatized displays
Rothstein Vitrinen

CATALOGUE

Concept
Prof. Dr. Peter Pachnicke, Wolfgang Volz, Jeanette Schmitz

Scientific consultant
Ulrich Köhler

Image legends
Dr. Thomas Bührke

Image sourcing
Susanne Pieth, Judith Schüller

Image sources
Cover photo: Cirrus nebula, photo:
Wisconsin Indiana Yale Observatory & NOAO, T.A. Rector.AURA, NSF
P. 110/111: M. Kornmesser, IAU
P. 113, 119, 121–123: Photos: Wolfgang Volz
P. 114–118: Photos: Thomas Wolf
P. 120: Photo: Parschik and Lüning

Layout
Volker Pecher, Essen

Proofreader
Thomas Machoczek

Production
Klartext Verlag, Essen,
Druckerei Uhl GmbH & Co. KG, Radolfzell

ISBN
978-3-8375-0257-2

Further information
www.dlr.de
www.gasometer.de

RUHR.2010 Kulturhauptstadt Europas

ORGANIZER

GASOMETER OBERHAUSEN

PARTNER

DLR Deutsches Zentrum für Luft- und Raumfahrt e.V. in der Helmholtz-Gemeinschaft

SPONSOR

evo

MEDIA PARTNERS

WDR 5 WAZ NRZ